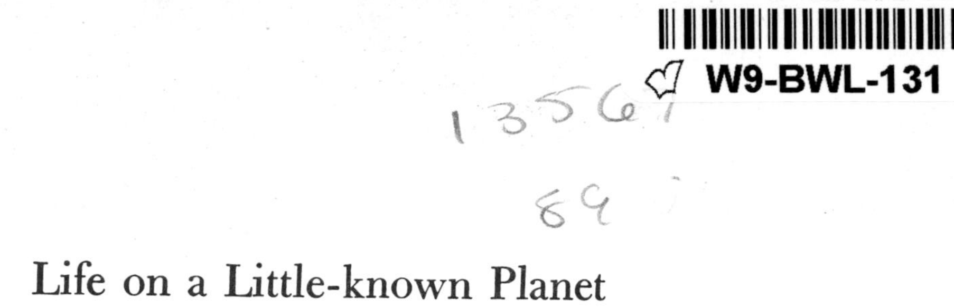

Life on a Little-known Planet

Howard Ensign Evans

Life on a Little-known Planet

With a new Foreword by

John Alcock

Illustrations by

Arnold Clapman

The University of Chicago Press
Chicago and London

Grateful ackowledgment is made for permission to use the following copyrighted material:

Don Marquis, *archy and mehitabel*
(Doubleday & Company, Inc.)

C. B. Williams, *Insect Migration*
(Wm. Collins Sons & Co. Ltd.)

W. M. Wheeler, *Foibles of Insects and Men*
(Alfred A. Knopf, Inc.)

W. M. Wheeler, "The Dry-Rot of Academic Biology," in *Science*, Vol. 57 (January 19, 1923).

Chapter 3 is an expansion of an essay which first appeared in *Harper's*, © 1966 by Harper's Magazine, Inc. Chapter 8 is an outgrowth of a review of Oldroyd's *The Natural History of Flies* published in *Scientific American*, © 1966 by Scientific American, Inc.

The University of Chicago Press, Chicago 60637
The University of Chicago Press, Ltd., London

University of Chicago Press edition 1984
Printed in the United States of America

93 92 91 90 89 88 87 5 4 3 2

Library of Congress Cataloging in Publication Data

Evans, Howard Ensign.
Life on a little-known planet.

Reprint. Originally published: New York: Dutton 1968.
Bibliography: p.
Includes index.
1. Insects—Behavior. I. Title.
QL496.E93 1984 595.7'051 84-86
ISBN 0-226-22258-6 (pbk.)

This book is dedicated to
the book lice and silverfish
that share my study with me.
May they find it digestible!

Contents

Illustrations

Foreword

When I was a graduate student immersed in the study of bird behavior, my only interest in insects was in the role they played as bird food. Luckily for me, my limited ornithological perspective was shattered the day I saw a film on the ground-nesting sand wasp made by Howard Evans. As Evans described the natural history of the wasp, I realized for the first time that some insects have lives every bit as intriguing as those of birds. This revelation set in motion a career change for me. Now I am first and foremost a watcher of insects and only secondarily a watcher of birds.

When you consider that there are several million species of insects, each wonderfully distinctive, and fewer than 10,000 species of birds, it is strange that there are millions of bird watchers but only thousands of insect enthusiasts. No doubt the relative size of birds and insects has something to do with it, but to learn to focus on small things does not require a great adjustment. Once the adjustment is made, a new world suddenly appears with a vast array of animals whose behavior is bizarre and fascinating.

Life on a Little-known Planet provides a wonderful antidote to the widespread notion that the "world of small and creeping things" may be ignored as insignificant and uninteresting. No one can read this book and come away without an improved awareness of the magnificent diversity of insects which simply swamps anything that the mammals, birds, and other vertebrates have to offer. Professor Evans promises that he will "try not to sound too much like a professor," and in this he succeeds admirably, entertaining us as he conveys his enthusiasm for insects. His clear and humorous prose ranges over the whole of entomology. Among the many things we learn about are the mating behavior of springtails, the neural basis for the ability of cockroaches to escape the flyswatter, the number of mosquito larvae that a dragonfly can consume during its juvenile life stage, the biochemistry of firefly light, the migrations of grasshoppers, and the lifestyles of flies associated with buried human corpses and the nasal septa of living human beings. Along the way we also learn something about the people who have studied these subjects, people who are sometimes as fascinating as the insects. Although nearly twenty years have passed since the original publication of the book, its contents have not lost their validity and charm. Nor is there anything dated about the author's contention that we ought to devote some of our energies as a society to fundamental research on the world around us, insects included. Thus the insect world and the human species have cause to rejoice at the reprinting of this delightful and instructive book, for we both have much to gain from its message.

It is too bad that books like *Life on a Little-known Planet* are not required reading for beginning biologists in high school and college, for this would surely open weary eyes and refresh minds numbed by endless requests to memorize the Krebs cycle or the molecular constituents of the gene. There is, Evans tells us here in lively fashion, an unknown universe just off our back porches, mysterious and extravagant, yet accessible and inviting, with a biology that is fun to learn about for amateur and professional alike. I am glad that we have the opportunity, through republication of this book, to join Howard Evans in admiring the silverfish and yellowjacket, the bedbug and the locust. We may always be a minority, but at least we will be in good company.

John Alcock

Preface to the 1984 Edition

Nearly two decades have passed since *Life on a Little-known Planet* was first published and times have changed. It is now clear to all but a few space enthusiasts that we shall never know of any life save that on earth. And it is equally clear that habitat destruction on earth is such that we shall never know many of the organisms that have resulted from that grandest of all dramas, organic evolution. This may prove the ultimate tragedy; that so much of the diversity, the genetic variability, the inspirational value of life on earth is already gone, or soon will be; that gone too is all hope of that peaceable kingdom of which we dream. Insects have (whether we like it or not) been the most abundant product of organic evolution. Of the estimated two or three million species on earth, we know of fewer than a million, and we know really very little about most of those. Yet we pour so much of our resources into extravagant life styles and the defense of those life styles that little is left for research on our fellow creatures, however essential some of that knowledge might someday prove to be.

But prophets of doom are not popular in the

headlong rush of our society, so I shall not persist in that vein. Rather let me say that discovery is one of the greatest joys, and it is still there to be had in abundance. Discovery may be on several planes. It may be personal, one's discovery of a fact of nature previously unknown to him. It may be practical, a new way of putting information to use in some human discipline. It may be universal, a fact new to science. It is out of discoveries, pure and simple, not used for personal and national aggrandizement, that the society of the future will be built, if we let it be. Since insects constitute the major class of our coinhabitants of planet Earth, it is good to become better acquainted with them. They provide a very rich source indeed of the joys that accompany discovery. This book is an effort to convince you that this is so.

Acknowledgments

Chapter 3 is an expansion of an essay which first appeared in *Harper's*, © 1966 by Harper's Magazine, Inc. Chapter 8 is an outgrowth of a review of Oldroyd's *The Natural History of Flies* published in *Scientific American*, © 1966 by Scientific American, Inc.

I am greatly indebted to various specialists who have reviewed individual chapters for errors of fact and interpretation. These are: Kenneth Christiansen (Chapter 2), Louis Roth (Chapter 3), George H. Bick (Chapter 4), Richard D. Alexander (Chapter 5), James E. Lloyd and John B. Buck (Chapter 6), Lincoln Brower and Jo Brewer (Chapter 7), Carroll M. Williams (Chapter 9), Ashley B. Gurney (Chapter 10), Stanley Flanders (Chapter 11), and Elwood C. Zimmerman (Chapter 12).

Robert Matthews and Mary Alice Evans read major portions of the manuscript critically, and Janice Matthews read the entire manuscript and made a great many helpful suggestions. Katherine Pearson typed the manuscript and also made many useful suggestions.

H. E. E.

The Universe as Seen from a Suburban Porch

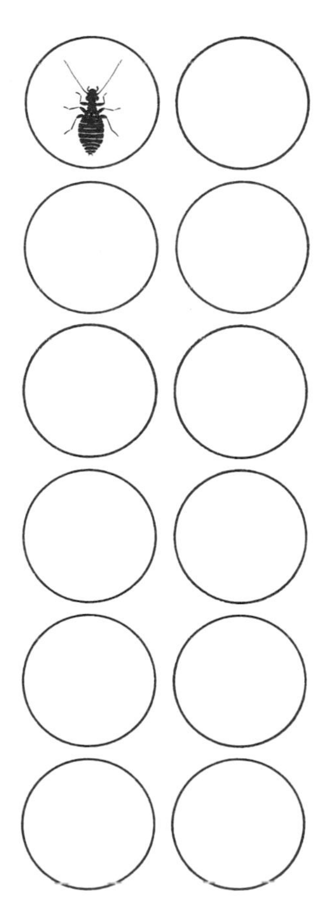

I begin this book on a summer's day on my back porch, as good a place to contemplate the universe as any, and better than some. The calling of blue jays and the rustle of leaves remind me from time to time of the richness of this our earth, and the drone of traffic on the nearby superhighway, accented now and then by the cries of children from neighboring houses, remind me of man's dominion over it. I warm my bones in the sun, mother sun, without whose energy no leaf would rustle, no traffic drone. The summer sun evokes a dreaminess unlike that of night; it is as though one's flesh and bones had deliquesced, had somehow slipped back for a moment into the elemental stuff from which they were long ago raised and sent upon their strange career. I seem to see a world of searing flame, of forms without form, of shrieking silence. Then a dog barks, and I know I have been asleep.

A fly pivots on the rail. Does he know he is trespassing on the sanctum sanctorum of man, the suburban home? Why does he appear so self-assured, so unmindful of me? Doesn't he know he was once a mere maggot, bred in a garbage pail, that as a fly he is bristly, coarse, and wholly unlovable? Evidently he concerns himself with none of these things, nor even with the fact that I, owner of this porch, represent the ultimate form of life on earth, the apogee of evolution, the image of God. Of course, after forty-nine years I do not have nearly so many offspring as he has fathered in a few short days, and I most certainly can't make a forward somersault and land on the ceiling. But I know all about him (well, quite a bit anyway), and he knows nothing about me. And my kind has learned to split the atom and to send vehicles into space. Knowledge is power; and we could reduce the earth to rubble if we wished (and even if we didn't wish). What has the fly done? (Come to think of it, this may not be a very fruitful line of thought. I have read somewhere that insects are several times as resistant to radiation as man, and I recall how flies revel in rubble and corpses.) I flick my hand, and the fly is gone. I am master of my thoughts and of my porch rail.

In the distance I can see the hills of Concord, where slightly more than a century ago there roamed a man with strange ideas in his head. Would he approve of this great new superhighway, roaring with machines carrying people to thousands of frantic destinies? (He would at least not have spoken of the *quiet* desperation of our lives!) Walden Pond is now the site of a white cement swimming pavilion, a trailer park, and an assortment of hamburger stands, and Thoreau's "clear and deep green well" is spangled with beer cans. Yet a fair number of people still pay homage to the site of his cabin, and most of them are young people. It is encouraging that his words still live in this very different world, that with courage it is still possible to listen to a different drummer. To Thoreau, man was a guest on a planet rich in texture, and filled with significance. We have learned a great deal since his time, but seem unable to conclude very much from all of it.

The fly is back. Well, if Thoreau could converse with a woodchuck, I can do so with a fly. He is a symbol of our times, a beast that is very much at home in our midst. He has developed resistance to DDT and to BHC, chlordane, dieldrin, and several other insecticides. He is not resistant to flypaper, but the stores no longer carry that symbol of nineteenth-century ineptitude. Odd that we can flatten mountains and land devices on the moon, yet do so little about the fly. Is it possible we don't know him well enough? Vincent Dethier, at Princeton, has been studying the behavior of the fly for much of his life, and the Australian entomologist A. J. Nicholson has studied fly populations in laboratory cages for many years. Neither claims to have fully grasped the fly's capacities. Is the fly a more intricate machine than he appears, or are we less clever than we suppose ourselves to be?

But of course we are clever. Just a few years ago (in terms of geologic time) we learned to use fire, and now we have mastered nuclear power, the ultimate form of energy. Just a few days ago we believed that our flat world ended in the Mare Incognitum, and now we study galaxies spinning in a space so vast that we have no language to describe it. Tomorrow we expect to put a man on the moon. Already we have made many wonderful photographs of the lunar landscape and have gathered data on the texture and chemistry of its surface. Evidently the moon is made of pretty much the same stuff as the earth, which is a pity—it would have been worth more as green cheese.

We are clever, too, in having probed so deeply into the physics and chemistry of life. In 1953 Harold C. Urey and Stanley L. Miller, at the University of Chicago, produced amino acids, the building blocks of proteins, by passing an electric charge through a mixture of methane, ammonia, hydrogen, and water, and since that time other, more complex molecules have been produced in the laboratory. Now Arthur Kornberg and his colleagues at Stanford University have caused an artificial virus to duplicate itself in a test tube. Our space program is in part predicated on the hope that we may sometime, somewhere, locate life in its early stages on another planet—or possibly find a planet where evolution has produced a whole gamut of unimaginable beings, some perhaps intelligent like us. A new science has been born: exobiology, the study of extraterrestrial life. For the moment it is a science without a subject matter, since we have not found life elsewhere in the universe —and there is a very good chance that we may never find it.

The moon we know to be a sterile and incredibly desolate place, and there seems little reason for putting man there except that "it is there." But of course our moon efforts provide a pilot project for exploration of the planets, and we have long been curious about those fellow travelers of our solar system. Already we have sent space vehicles close to Mars and Venus, and the Russians have landed instruments on Venus. We now know that the surface temperature of our closest neighbor, Venus, is a little cooler than we supposed, a mere 536 degrees Fahrenheit—still hot enough to maintain pools of molten metal. And there is hardly any oxygen or water vapor in the atmosphere. All in all, Russia's Venus IV and our own Mariner V have confirmed an earlier judgment by Carl Sagan, of Harvard University and the Smithsonian Astrophysical Observatory—and as ardent a space enthusiast as you are likely to find. "Venus," remarked Sagan, "is very much like hell."

What about Mars, a much cooler planet that in our telescopes seems to show greenish flushes and evidence of "canals"? Judging from the photographs made by Mariner IV, its surface looks rather like that of the moon. And we know that Mars has little atmosphere, with almost no free oxygen and little water. The Martian polar caps, so conspicuous in our telescopes, we now believe to be made of "dry ice," that is, frozen carbon dioxide. It still seems possible that there may be simple plants on Mars, plants able to survive with little oxygen or water and capable of enduring the very cold nights and the sterilizing effect of the ultraviolet radiation that penetrates the scanty atmosphere. But I doubt it. No less a person than Philip Abelson, editor of *Science,* and a geochemist who has performed experiments bearing on the origin of life, commented recently that "the probability that life has originated on Mars is trivial."

It is universally agreed that Mercury is much too hot for life and has no atmosphere, while Saturn and the planets beyond are very much too cold and have atmospheres composed chiefly of ammonia and other poisonous gases. Jupiter, our largest planet, is believed to have a temperature of something like minus 200 degrees Fahrenheit, but it is shrouded in clouds, and there is a possibility that the heavy cloud cover produces a "greenhouse effect" on the surface and that somewhere in the ammonic mists some primitive creature is aborning. But again, I doubt it. At any rate I shall not hold my breath, at least not until we get to the moon. Jupiter, at its closest, is more than ten times as far from the earth as Mars, and more than one thousand times as far as the moon.

There are, of course, other systems of planets in the universe. Harlow Shapley, in his book *The View from a Distant Star,* speculates that so vast is the universe that there may be ten billion planets suitable for life. The trouble is that the nearest stars are some 60 trillion miles away—something like 200 million times as far as the moon, if I haven't lost a few zeros somewhere along the way. Of course, we don't really know if those nearest stars have habitable planets, and we have no way of finding out without going there (unless, of course, there are beings capable of sending messages through space). Yet there are people who in all seriousness speak of man populating other planets; and people who with a good deal less seriousness (I hope) speak of "wild-shot" space ships as a way of disposing of our excess population. An expedition to a "nearby" star might indeed find no habitable planets, and have to fly another several trillion miles to a second star. Such an expedition would resemble, in the words of Garrett Hardin, professor of biology at the University of California at Santa Barbara, a "latter-day interstellar Flying Dutchman." But the technological, psychological, and sociological difficulties in undertaking such flights are so fantastic that all this seems no more than another dream engendered by the summer sun.

On the whole I find my back porch a rather more comfortable bit of the universe, my fly a good deal more likable than he seemed at first. He doesn't know it, but some of his cousins have been trained as astrobugs. Biosatellite II successfully orbited the earth with an assortment of creatures that were supposed to teach us something about growth and reproduction under weightless conditions. According to the United Press, the passenger list included "10,000 vinegar gnats, 1,000 flour beetles, 560 wasps, 120 frog eggs, 875 amoebae, 13,000 bacteria, 78 wheat seedlings, 9 pepper plants, 10 million spores of orange bread mold, and 64 blue wild flowers." Tsk, tsk, not a single yellow-headed pickleworm, the species that surely holds the secret to life in space.

For all my sarcasm, I would not for a moment suggest that we drop our space program or even decrease its budget. Space exploration is an inevitable step in the growth of human culture. Despite the problems and the multibillion dollar price tag, it has been a much healthier occupation than some other human exploits. At the same time I think we should label it properly not as science (at least not as biology) but as sport. "Gemini home with seven records" reads a newspaper headline, as though it were reporting a track meet. As sport it is a convenient escape valve from our truly pressing problems here on earth, and a benign outlet for international tensions. Obviously, the science of astronomy stands to gain a great deal as we learn more about the planets, and especially if we are able to establish observatories outside the earth's atmosphere. It is also true that there has been some "spin-off" from the space program into terrestrial biology; for example, advances in electronic miniaturization and automation have been found useful in studies of animal navigation. But many students of life would agree with G. G. Simpson's remarks in his book *This View of Life:*

"Let us face up to the fact that [the search for extraterrestrial life] is a gamble at the most adverse odds in history. Then if we want to go on gambling, we will at least recognize that what we are doing resembles a wild spree more than a sober scientific program.

"To some it seems that the reward could be so great that facing any odds whatever is justified. The biological reward, if any, would be a little more knowledge of life. But we already have life, known, real and present right here in ourselves and all around us. We are only beginning to understand it. We can learn more from it than from any number of hypothetical Martian microbes. We can, indeed, learn more about possible extraterrestrial life by studying the systematics and evolution of earthly organisms. Knowledge from enlarged programs in those fields is not a gamble because profit is sure.

"My plea then is simply this: that we invest just a bit more of our money and manpower, say one-tenth of that now being gambled on the expanding space program, for this sure profit."

How much do we know about life on this little-known planet beneath our feet, the planet earth? We have not even approached the end of cataloguing the creatures that share the earth with us: and this should be the very first step in our knowledge. At the present pace it is probable that many will become extinct as a result of human expansion before we even knew they were there. Is it sensible to poke about for strange beings in space while we blindly exterminate those about us before we have even made their acquaintance? It is said that not much more than half the organisms on earth have yet been described. As one who has several times discovered insects new to science in his own back yard, I can believe this. And once new species have been described and placed in a museum, what do we really know about them? Do we know what role they play in nature, how they affect other species, how they evolved? Do we know how they behave in various facets of their life history, and do we know what makes them do the things they do? Do we know all the subtle ways they impinge upon man? Do we hope to understand man apart from the world in which he evolved?

More and more I am talking myself into a genuine friendship with the fly grooming himself on my porch rail. How different would the world be without the fly to help in the decomposition of wastes and carcasses? Without the fly as an experimental animal, how much less would we know about population cycles, about nervous function, about heredity? What is a fly worth, an oak tree, a crow, a wisp of thistledown? By how much would life be diminished if Shelley had not written his *Ode to a Skylark*, if Emerson had not penned *The Rhodora* or Lanier *The Marshes of Glynn*? How many persons would not be alive today if we had not discovered penicillin, the improbable product of a lowly green mold? If it is true that half our new drugs are being produced from botanical sources, how can we afford to neglect or destroy any portion of the earth's green mantle? Who can say what obscure plant or animal may someday be precious to us? Are not all precious, since in fact we understand so little about the interdependence of living things, since life itself is the most precious thing of all? The earth has spawned such a diversity of remarkable creatures that I sometimes wonder why we do not all live in a state of perpetual awe and astonishment.

A suburban back yard may not be as good as a tropical rain forest for demonstrating the diversity of life, but in the final analysis the eyes one sees with are more important than the setting. My neighbor is engrossed in his newspaper, and sees nothing but the latest case histories of human behavior—a subject of much interest, of course, for people are even more fascinating than flies. My own porch is, however, a more versatile earth probe, with an eye to the fly and a moment for the caterpillar swaying from the end of a silken thread suspended from an oak tree. To the puzzlement of my neighbors, I have set up a Malaise trap in my back yard—a tent-like insect trap invented by the Swedish entomologist René Malaise. Even here, surrounded by lawns and houses, the trap harvests thousands of insects in the course of a summer, insects that were simply flying along and blundered into it, for the trap is unbaited. Last summer I collected several wasps for the first time in Massachusetts, a second record from the United States for an introduced European cuckoo wasp, and at least one species new to science. (I have analyzed only the wasps, my specialty; there may well also be exciting records among the other insects.) By using traps baited with carrion or with certain volatile oils, I could undoubtedly attract certain kinds of insects from a wide area; and by replacing the yellow porch light with a blue one I could draw in great numbers of insects on warm evenings. The late Professor E. O. Essig, of the University of California at Berkeley, used to put out pans of water, and he found this an excellent way to collect flying plant lice, which were his specialty. Essig painted the pans various colors, and found that far more plant lice were captured in pans painted with "Glidden's carnival yellow." Why they prefer this color is anybody's guess; I doubt if plant lice know Mr. Glidden or have a natural yen for carnivals.

This is by no means the end of the devices one may use to inveigle insects into stopping by. One time-honored method for collecting certain moths and beetles is "sugaring," that is, smearing a mixture of stale beer, brown sugar, and crushed bananas on tree trunks. An ordinary tin can sunk in the ground to its rim, and containing a preservative fluid in the bottom and perhaps some bait suspended from the top, is likely to collect a wide variety of insects, spiders, and other creatures that crawl over the ground. Of course, all these methods are more productive in an undeveloped area, and one could be rather unpopular if he employed all of them in a small suburban yard. But the point is that one could, even in such a place, harvest more of the diversity of life than he could find time to study. This is especially true if one included the great many small organisms that flourish in the soil—the subject of Chapter 2. "I had no idea that there was so much going on in Heywood's meadow," Thoreau once remarked. One but needs the eyes to see it.

On a warm summer day the air is filled to a considerable height with small organisms that drift about in the air currents. Some thirty years ago, A. C. Hardy and P. S. Milne, of the University College of Hull, England, conceived the idea of sending up an insect net attached to a box kite, the net devised to open and close at specific heights, and thus sample the insects at known altitudes. Over the playing fields of the college they collected 839 insects at heights of from 150 to 2,000 feet in 125 hours of flying time. All of them were small, light-bodied insects with weak powers of flight, such as plant lice and small flies and parasitic wasps. At about the same time P. A. Glick, of the United States Department of Agriculture, was collecting insects in special traps fitted to airplanes over Tallulah, Louisiana. He collected 30,033 specimens at altitudes of from 20 to 15,000 feet. Again, the kinds taken were mostly light-bodied forms. Surprisingly, wingless insects and mites were found to occur high in the air, and a spider was captured at the highest altitude at which collections were made. Estimates showed that a column of air one mile square, starting 50 feet from the ground and extending to 14,000 feet, contained an average of 25,000,000 insects. Such a column also contains vast numbers of pollen grains, spores, seeds, bacteria, and other small living things. We are only just beginning to appreciate the importance of the "aerial plankton" in the distribution of living things, including human allergens.

The earth is a planet so richly endowed with life that we shall never run out of problems worthy of study. Each of the two or three million species of animals and plants (we can't estimate the number any better than that!) is unique, and fills its role in nature in its own individual manner. More often than not we know little or nothing about that role; and so complex is the web of nature that such answers as we have are generally partial and uncertain. Biology has been called the science of the future. I agree, and would add that the sooner we appreciate the richness of life on earth, and the almost certain absence of life on other planets within our reach, the sooner we can get on with it.

I have been speaking about what "we" know about life on earth. But who is "we"? Having been trained in entomology, I know a good deal about certain insects, but almost nothing about whales, and I am distressed that most species of whales are becoming extinct and that neither I nor my children will ever know very much about them. The fact is that "we," meaning mankind in general, know so little about living things that we scarcely appreciate the drama that is there, the exciting frontiers pushed into vast jungles of ignorance. Each probe into the jungle is manned by only a few lucky (and very busy) individuals, and, sad to say, there are hardly any good reporters there to convey the excitement back to the rest of us. When we speak of human knowledge we need to remember that much of it is compartmentalized and the province of a select few—even though it may be utterly fascinating and even potentially revolutionary from a human point of view.

As a museum curator and research entomologist, I can hope to report only on a small segment of what is known in that particular compartment of knowledge pertaining to insects—and to suggest the depths of the forest ahead. I could equally well be a student of worms, of fungi, or of sea squirts; or an explorer in any of the wonderful blocks of jungle between biology and chemistry. But I am not. As a matter of fact, few groups are better suited to demonstrate how little we know about our planet than the insects—how little the experts know, and how very little of this knowledge has reached "the man on the street." In variety of different kinds, insects exceed all other living things. In numbers of individuals they are exceeded only by certain plants and microorganisms, and, in the sea, by crustaceans. (One estimate puts the total insect population on earth at any one time at a billion billion.) In the past, insects have helped bring about the fall of civilizations, and even today they cause widespread distress and the expenditure of billions of dollars. Yet such is our culture that we teach our children almost nothing about them!

Admittedly I am prejudiced. As a farm boy in the Connecticut Valley, I grew up with bean beetles and corn borers. My first job was picking off tobacco hornworms for a penny each, and my first wealth (several dollars!) came when I sold to a neighborhood hobbyist some of the many moths that came to the lights of our roadside stand. Happiness to me was pursuing a butterfly across a meadow splashed with goldenrod and joe-pye-weed. As an early teen-ager, I helped to found the Hockanum Nature Club, which operated a museum occupying an old woodshed and possessing an assortment of treasures such as birds' eggs, pressed leaves and wildflowers, and a great variety of local insects.

There were, of course, many efforts to discourage me from continuing this sort of thing, and they had largely succeeded when I came in contact with a very unusual teacher: J. A. Manter, of the University of Connecticut. In his quiet way Professor Manter introduced me to the world of professional entomology, and from that moment on insects have been my life. It was he who guided my senior problem on the insects of fallen timber (that was just after the 1938 hurricane), and it was a preoccupation with the parasites of these insects that eventually led to my specializing on wasps in graduate school. Even now, my wife and children claim they play "second fiddle" to my wasps. My usual answer is that wasps are, after all, elegant creatures, and they are lucky to be playing second fiddle to them and not, let us say, to chicken lice.

I have already told of our Ithaca home and some of the wasps that shared it (in *Wasp Farm*, 1963). The present book is mostly about other kinds of insects that have interested me, though I may sneak in a few words on wasps before I am through. I shall try not to sound too much like a professor or museum curator (though my nearly twenty years in these roles may show through now and then). Sometimes it seems desirable to slip in a bit of technical information on classification and terminology, but I have put most of this at the end of the book, where the reader may ignore it if he wishes. I have also listed there, according to chapter, a number of references which a reader whose curiosity is aroused may consult for further information.

When talking about insects, it is sometimes necessary to use their scientific names. There are so many kinds that people have never invented common names for all of them. Think of the problems in concocting so many names. There are nearly 300,000 kinds of beetles alone, and that is more than the number of entries in the unabridged dictionary on my desk. I have used common names wherever they exist, but where they don't exist I have not hesitated to use the scientific names. They provide the only international language we have in an era when such a language is much needed. In scientific names the group name (generic name) comes first and is capitalized, while the species name comes second and is not capitalized. Thus one of the leading characters in Chapter 6 is a firefly called Photinus pyralis, a very attractive name that happens to mean "a fiery little light." There are many closely related fireflies also bearing the generic name Photinus: for example, Photinus scintillans, Photinus ignitus, and Photinus floridanus—all euphonious names with meanings that are easy to guess. Scientific names are a bit like Italian opera; the words are attractive and even exciting, but when translated they sometimes turn out to be a bit sillier or less expressive than we thought they were. Some assistance in pronouncing the scientific names I have used is also provided at the end of the book.

Entomologists are supposed to study insects, but in fact they often become involved with related organisms having more than six legs and thus not true insects: things such as mites, spiders, and centipedes. This is perhaps fair enough, since the word is based on *entomos,* which the Greeks used for a variety of crawling and flying creatures without bothering to count their legs. Incidentally, entomologists are always delighted to find people who can spell the name of their profession properly. I have always especially resented being called an antomologist, when I specialize on wasps, not ants. Antimologist is even worse, since it implies I am against -ologies, which I usually am not. An etamologist, I suppose, is one who has just eaten a scientist.

Anyway, this book is devoted to creatures with six legs, leaving the spiders and centipedes to the araneologists and the myriopodists. It would be nice if we didn't need these words at all. Living things are all of a piece, though it is such an exceedingly complicated piece that specialists have to compartmentalize one way or another. Another way to do it is to be a specialist on, say, genetics or behavior, and to use whatever animals are useful. Most of us cut the cake both ways, then cut it both ways again and again, until we finally write a long, tedious scientific treatise about a crumb. This is acceptable—in fact inevitable—but some of us tend to forget the cake of which it is a part. We can hardly blame the nonbiologist if he, too, doesn't see the cake.

Already this is beginning to sound like a lecture (or perhaps a sermon), so let me conclude quickly. The universe as seen from my porch consists mostly of insects. And I defy the science-fiction writers to dream

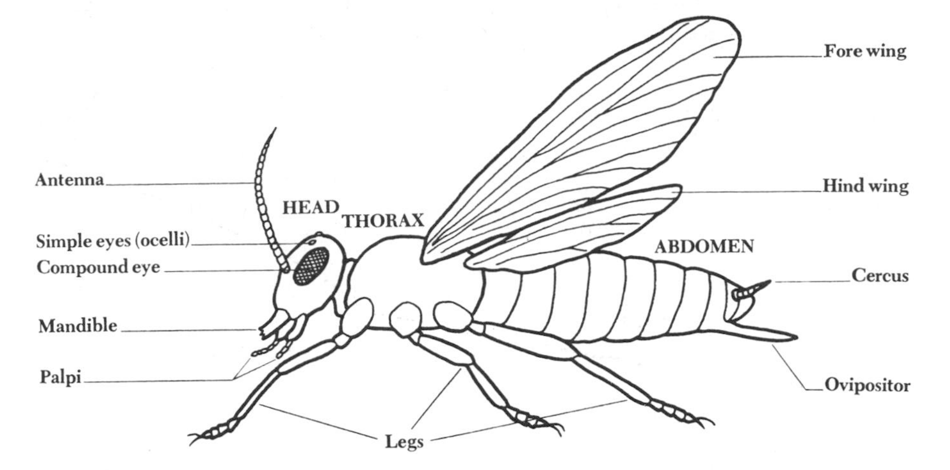

The major features of an insect. The wings are not modified limbs (as in birds and bats), but entirely new structures arising from the back—a feature shared only with angels.

up anything quite so wonderful as the mosquito that has finally persuaded me to go inside and try to reassure myself that I am master—at least of my own blood.

Notes on Some of the Terms Used

This seems to be a good place to make sure the reader understands what I mean by certain words that I shall use regularly. When I say "organism," for example, I refer to any kind of living thing. An "animal" is an organism capable of moving about and living (directly or indirectly) at the expense of plants, which are largely sedentary and possess the green pigment chlorophyll, which represents the link of all living things with the ultimate source of energy, sunlight. The use of the word "animal" to mean "mammal," that is, a backboned animal that possesses hair and suckles its young, is simply not correct. An animal is any living thing that is not a plant. (There are indeed quite a few microscopic plant-animals and things that are not quite either plants or animals, but fortunately they don't fall within the scope of this book.)

The "arthropods" form the largest of all groups of animals. Arthropods have jointed feet (which is what the word means: if you have *arthr*itis you have pains in the joints, and if you go to a *pod*iatrist it is because you have foot trouble). Arthropods have no backbone, but they are covered with an external skeleton to which their muscles attach. An insect (somewhat loosely speaking) is an arthropod with only three pairs of legs.

What is a bug? To an Englishman it is a bedbug (its original meaning). To an entomologist it is the bedbug and all his immediate relatives, that is, all insects with a specialized type of sucking beak (such as the stinkbugs, the chinch bugs, and even such offbeat creatures as water striders). (Some of the true bugs will be the star performers in Chapter 9.) To the layman a bug is any small creeping or flying animal, or even an organism one can't see, such as a virus or bacterium. If you develop a cold while visiting an Englishman, it is best not to tell him you have "picked up a bug." The word "bug," used by itself, is so ambiguous as to be of little use to anyone, and this is why entomologists refer to the bedbug and his relatives as "true bugs."

The true bugs make up a well-defined grouping of insects called an "order," in this case the order Hemiptera. Other orders of insects are the beetles (Coleoptera), the moths and butterflies (Lepidoptera), the flies (Diptera), and so forth. Each order is divided into families: for example, the fireflies make up a family of beetles called the Lampyridae, the plant lice (or aphids) a family of Hemiptera called the Aphididae. Families are divided into genera, and the generic name makes up the first part of the scientific name, as already explained. It is perfectly possible to enjoy insects without all this taxonomic folderol, but when one gets deeply involved with them it is necessary to have some sort of framework on which to hang the information he gathers, and it is pretty important that a worker in Russia, for example, uses the same framework I do. If the insects knew the names we called them, they would probably be crushed by their weight, and the perfect insect control would have been developed.

Cities in the Soil: The World of Springtails

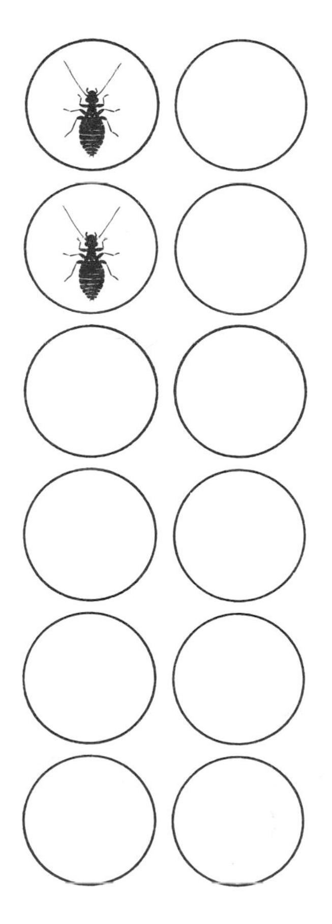

Man is a giant of the earth. Not only is he physically larger than all but a very few other animals, but his cultural heritage renders him all-powerful, and his horizons are not the very limited home territories of most animals, but the limits of the universe itself. It is odd that man is basically a lonely creature, speaking often of the brotherhood of man to man and of man and nature, but living nonetheless largely within himself. He may hear of the decimation of wildlife in Africa or of famine in India, but he sighs and goes about his work and his pleasures: never quite the animal he was or the god he would be.

The physical proportions of man provide him with a major handicap, for larger creatures require that much more space in which to live and that much more food to eat. It is unlikely that the earth can support more than about ten times as many people as it does now, that is, 40 to 50 billion individuals, and it is not at all certain that it will support that many. But a few acres will support that many insects, mites, and other small arthropods, and it is said that a square yard of good soil may contain more than 10 trillion bacteria and other microorganisms. The smaller the organism, the less space and food it requires and the faster it is able to reproduce its kind. If one makes a cross section of living things almost anywhere on land or sea, he finds a "pyramid of numbers," that is, a very great number of very tiny organisms, a slightly smaller number of slightly larger ones, a still smaller number of still larger ones, and so forth: each level of the step pyramid depending very largely on the one below it. On the top stands man and a very few other truly large organisms, and in a sense man's brain places him in a lonely and well-fed eminence above even the largest. The tragedy is that we on the top are so large that, even with our remarkable brains, we have difficulties understanding the creatures on the bottom—and very little time for this sort of thing—though without them the whole edifice would come tumbling. The pyramid of numbers is also a pyramid of knowledge (or ignorance), for what we know about an organism is often roughly proportional to its size.

In fact, what are the most abundant animals on earth? Like many simple questions, this is not an easy one to answer. First of all, does one mean the earth as a whole or merely the somewhat more than one-fourth of its surface that is above water? The oceans, after all, teem with quantities of living things that we can scarcely begin to measure, especially small crustaceans, the "brit" and "krill" of the sea. According to

The pyramid of numbers—which may also be thought of as a food pyramid, since the large organisms on the top strata are dependent upon those below; or a pyramid of ignorance, since we on top understand the lower strata so poorly.

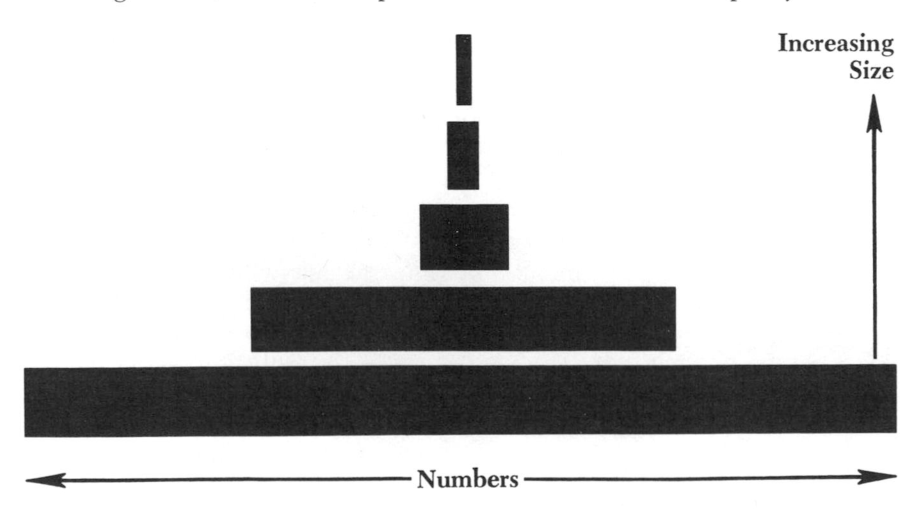

the oceanographer Sir Alister Hardy, members of one group alone, the copepods, exceed in abundance all other animals combined, not excluding the million or so species of insects, hardly any of which have ventured into the ocean. Nearly transparent and usually smaller than a grain of rice, the copepods are enormously important in the economy of the sea, and may become of much more direct importance to man as we learn to understand and to utilize more fully the resources of the sea. These copepods must themselves feed on something, of course: they graze the vast "meadows of the sea," consisting mainly of microscopic plants called diatoms, which, like land plants, utilize energy from the sun to convert simple substances into the complex organic compounds upon which all animals depend. It is said that at times the water beneath one square meter (slightly more than a square yard) of the Gulf of Maine may contain 7 to 8 billion diatoms, and that the Gulf of Maine as a whole may contain 4 million tons of those exceedingly small and almost weightless crustaceans, the copepods.

What of the land? Just as we can gaze out to sea and be aware of nothing but waves, sky, and a sea gull or two, so we can stand in a field or walk a golf course and have not the slightest notion—and usually not the slightest care—of the world beneath our feet. It is a matter of focus. Peter Farb prefaces his attractive little book *Living Earth* with the following remarks:

"We live on the rooftops of a hidden world. Beneath the soil surface lies a land of fascination, and also of mysteries, for much of man's wonder about life itself has been connected with the soil. It is populated by strange creatures who have found ways to survive in a world without sunlight, an empire whose boundaries are fixed by earthen walls. . . .

"The apparent lifelessness of a piece of earth is an illusion. Specks of protoplasm swarm through the soil granules, invisible to the naked eye. Larger animals and plants are linked to them by invisible threads of food supply and living conditions. Similarly, there are interlocking relationships between the larger creatures, and so up the pyramid until we come to its apex, the fabric of Life itself."

Disregarding microorganisms and nematodes (for even an entomologist is guilty of ignoring the lowest steps of the pyramid), we can say that arthropods make up most of the population of these little-known cities in the soil, just as very different arthropods (crustaceans) make up much of the fauna of the seas. The soil fauna contains a remarkable diversity of animals: virtually every major group is represented, except for several groups occurring only in the ocean. However, in most areas two types of animals predominate: the mites, which are eight-legged as adults and therefore grouped with the spiders, and the springtails, which are six-legged and therefore usually considered insects.

One of the early and now classic censuses of soil animals was made by George Salt and his associates at the University of Cambridge in 1943. Using the best available techniques for taking soil samples and for extracting the arthropods from them, these workers undertook to estimate the numbers in an ordinary English pasture. The following is their calculation of the soil population of one acre of pasture soil on a day in November:

Mites	666,300,000
Springtails	248,375,000
Root aphids and other sucking bugs	71,850,000
Bristle-tails (a group of wingless insects)	26,775,000
Various kinds of centipedes and millipedes	22,475,000
Beetles	17,825,000
Various other arthropods	15,200,000

In this particular acre they estimated a total of well over a billion arthropods. Their richest samples contained about 50 per cubic inch. Surprisingly, similar surveys of pastureland in both East Africa and West Africa provided very similar proportions of mites and springtails: in each case the two together made up 80 to 90 per cent of the total, with mites more abundant than springtails, exactly as in England.

Not only do these two groups predominate in tropical and temperate regions, but even in areas close to the poles. On the Antarctic continent they are among the very few permanent inhabitants, occurring in abundance wherever there is bare soil. Both springtails and mites have been taken as far south as the 84th latitude and as high as 6,000-feet elevation in the Antarctic mountains. Here they are sometimes subjected to temperatures as low as −85° Fahrenheit for long periods, but they remain dormant and revive when the dark rocks where they live absorb enough heat to enable them to feed and reproduce again.

Forested areas often contain even larger numbers of soil arthropods than open country, especially if there is deep leaf litter and humus to hold moisture and to provide numerous air spaces. On the other hand, clay soil has fewer and smaller air spaces, and sandy soil permits the moisture to percolate through, so these soils are usually relatively sterile. William Morton Wheeler, of Harvard University, used to claim that ants were the most abundant of all terrestrial arthropods—Wheeler was, of course, an authority on ants, and therefore highly attuned to their presence. In fact, in some sandy and clay soils, and in arid and semiarid areas generally, ants may very well predominate. Ants, of course, average a good deal larger than springtails and mites, and belong to a slightly higher level in the pyramid of numbers, so one should not expect them to be as abundant generally. The arthropods taken by Professor

Salt and his coworkers from an English pasture contained ants in a proportion of less than half of one per cent.

Under certain local conditions, springtails may exceed even the mites in abundance. A study of the fauna of grass tussocks in England showed springtails to make up nearly two-thirds of the arthropods, with 96.5 per cent of the springtails consisting of one species. Study of an onion field in Egypt also revealed a majority of springtails as compared to mites. Unfortunately, a good deal depends on the techniques of extraction one uses, whether or not he happens to hit a population peak of some particular mite or springtail, and so forth. Kenneth Christiansen, of Grinnell College, Iowa, who is one of our leading American authorities on springtails, has surveyed published accounts of springtail abundance and found tremendous variation. He goes on to say that he feels that most estimates are too low, since his own extensive hand sorting of samples of Iowa farm soil has led to figures as high as 100 million springtails per square meter: in other words something like 400 billion per acre. Professor Christiansen admits that this may represent an upper limit, rather than an average population size, but as one who has sometimes seen springtails in incredible numbers in recently manured farmland, I for one am not inclined to argue with him. In fact, considering the variables involved in the study of soil populations, I am not inclined to start any arguments at all. It seems to me perfectly safe to say that springtails are by far the most abundant six-legged animals on the land masses of the earth, and the most abundant of all animals on land having any legs at all with the probable exception of the mites. He who would prove me wrong had better start counting!

Despite their abundance, springtails apparently escaped the attention of naturalists of ancient times. Any real knowledge of springtails necessarily followed the first use of lens and microscopes, for only a few springtails exceed a quarter of an inch in length. A great stride in the study of soil animals was made by the Italian entomologist Antonio Berlese, who invented a device for extracting these animals from soil and litter, the so-called Berlese funnel. This apparatus, in its simplest form, can be made and used by almost anyone, and he will be rewarded with an array of animals a good deal stranger than those to be found in books on prehistoric life: in fact, these are for the most part the descendants of very ancient animals. A Berlese funnel is simply a large funnel with a wire mesh placed across it near the widest part. The sample of soil or leaf mold is placed on top of the wire, and a collection jar is placed beneath the funnel. A source of heat is then placed above the soil sample: an ordinary light bulb is fine. The heat and resulting dryness cause most of the soil animals to descend and eventually (over a few days) to fall into the jar, where they can be collected in alcohol or kept

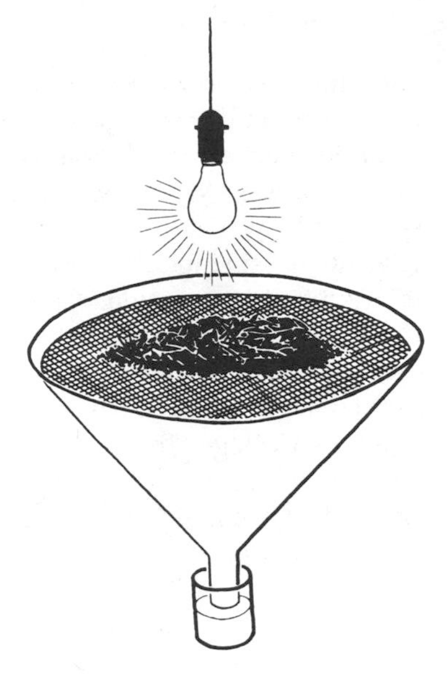

A Berlese funnel. Much more sophisticated devices for extracting insects from the soil are now available, but none is as easy to make or to use as this simple apparatus.

alive, but much concentrated, in a little damp leaf mold. Various modifications of this technique, as well as very different flotation techniques, are described by D. Keith McE. Kevan, of McGill University, Montreal, in his excellent book *Soil Animals.*

It was less than a century ago—in 1873, to be exact—that the first major publication on springtails in English appeared. It was the work of that remarkable amateur zoologist Sir John Lubbock, later Lord Avebury. Lubbock, a banker by profession, lived near Charles Darwin at Down in Kent, and was a close friend of Darwin, who was his elder by twenty-five years. Lubbock's first scientific work, published when he was only nineteen, concerned some of Darwin's collections, and he also made many of the illustrations for Darwin's great work on barnacles. Lubbock became one of the most prolific writers of his age, his contributions covering various aspects of zoology, botany, geology, and anthropology, and including several very popular books with such titles as *Beauties of Nature* and *Peace and Happiness.* As a member of Parliament, Lubbock did much to shorten the hours of labor, to promote the study of science in the schools, and to preserve natural areas as well as historically important sites such as Stonehenge. Although he never went to college himself, Lubbock undertook a rigorous self-education, and at thirty-seven he was made vice-chancellor of the University of London. Later he became a trustee of the British Museum, and in the course of his life he was a member of virtually every scientific society of his time, and at one time or another president of most of them.

Busy as he was, Lubbock nevertheless found time to study various minute animals that he found in his own garden and elsewhere. It was in his garden that he discovered Pauropus huxleyi, a peculiar little creature not quite a millipede and not quite a centipede, in fact forming a new class of arthropods, the Pauropoda. Lubbock was the first to recognize the springtails as a group separate from other wingless insects, and for them he coined the word by which they are still known: Collembola, from the Greek words for "glue-peg," with reference to the so-called adhesive organ, or "ventral tube," near the middle of their body on the lower side. The excellent colored illustrations in Lubbock's monograph, made by the deaf-mute artist A. T. Hollick, called attention to the remarkable colors of some springtails. Lubbock began his book with the following comments:

"The Collembola . . . have hitherto been but little studied in this country. Yet if a fallen bough be examined, a heap of moss shaken over a pocket-handkerchief, or any long herbage swept with a hand-net, the naturalist will not fail to find, together with numerous beetles, flies, and other insects, certain delicate, hexapod, active little creatures; the majority of which will endeavour to escape not only by running with agility, but also springing with considerable force, by means of a sub-abdominal, forked organ, which, commencing near the posterior end of the body, reaches forward, in most cases, almost as far as the thorax. . . . Some species . . . live on the surface of ponds; some are found on the sea-shore; others . . . in houses; but the majority of species frequent fungi, decaying leaves, moss, or loose soil: in fact, wherever there is any decaying vegetable matter, Collembola may be found in abundance."

Lubbock went on to review what was known in his time of the structure and classification of springtails, along with the few available facts of their life histories. He was shrewd enough to remark that "we cannot, I think, regard them . . . in the strictest sense, as true insects": a view which many contemporary zoologists accept. True, springtails have six true legs, simple antennae, and eyes (except for the many blind forms) and mouthparts not unlike those of some primitive, wingless insects. On the other hand, the so-called Malpighian tubules, the "kidneys" of insects, are wholly lacking, and the type of cleavage of the egg is different from that of true insects. Even more strikingly, the abdomen of springtails has only six segments as compared to eleven in all true insects. Then there are the several unique structures on the abdomen: the spring, the "catch" that holds the spring when not in use, and finally the ventral tube, or "glue-peg," which gives the Collembola their name. The use of the ventral tube for adhering to the surface can best be observed when the springtail is resting on a wet glass surface. However, most workers believe that the primary function of the ventral tube is for absorbing water. Springtails are extremely sensitive to drying, and specimens in captivity have been observed to extend their ventral tube into droplets of moisture and withdraw it. When a colored stain is added to the moisture in the container, the stain enters the body through the ventral tube, more rapidly so in a drier atmosphere. Also, when the ventral tube is punctured, the springtails perish rapidly except when the atmosphere is saturated with water vapor. Some springtails are able to extend the ventral tube into a long finger that is used to transfer water to the mouth or even to clean themselves. Others have been seen to rub their legs and mandibles against the tip of the ventral tube, and it seems possible that the tube is the source of a fluid used for cleaning or lubricating these parts.

The spring and its catch are also remarkable structures. The spring is best developed in species living above the soil surface at least part of the time, and it is doubtless a very effective escape mechanism. It is operated by powerful muscles that work synchronously with smaller muscles that release the catch, thus sending the springtail flying into space. The larger species may jump six inches or even more; as one observer has remarked, "It is as if a man could cover a mile in nine or ten bounds." The catch is located just in front of the spring, and is a small projection with two sets of teeth that hook on to the inner sides of the two prongs of the spring, like the man who is said to have "used a wart on the back of his neck for a collar button." The source of these two colorful quotations, by the way, is another nonprofessional biologist who was much attracted to the study of springtails: Charles Macnamara, of Arnprior, Ontario.

Perhaps the most conclusive evidence that these remarkable animals should not be regarded as insects is of a more indirect nature. There is very good reason for believing that the insects evolved from some primitive many-legged creature, a "centipede" in the broad sense of that term. Some of the near "missing links" are still with us, and can often be found in the very same soil samples with springtails: symphylids, diplurans, and bristletails, creatures common enough in certain places, but well known only to specialists. The bristletails do have "domesticated" members, the so-called firebrats and silverfish. The latter are very much at home in my house, and have even been known to take a bath with me. I don't object seriously; they are clean animals ("paper-eaters," the children call them)—and after all, it is not often that one is privileged to entertain a not so missing link, especially in his bathtub.

The point is, we do know (or at least think we know) some of the stages the insects must have passed through in their rise from many-legged creatures, but the springtails simply are not on or near the trunk of this genealogical tree. They appear to represent a separate and inde-

(a)

(b)

(c)

Three kinds of springtails. The species that occur on top of the soil often have long antennae and springs and fairly good eyes (*a*). Those living in the upper layers of the soil tend to have shorter antennae and springs (*b*), while those living deep in the soil are generally white in color, blind, and have short antennae and no springs at all (*c*).

pendent experiment in six-leggedness, good enough to thrive in abundance in protected places on earth, but a little too much like an invention of Rube Goldberg ever to develop into anything else.

Actually, springtails do assume a good many different body forms, depending in large part on their habitat. Those occurring deep in the soil tend to be very small, white in color, blind, and with short antennae and legs and usually no spring at all. Those occurring in upper soil strata tend to be a little larger, to have more color in their bodies, and to have longer appendages. Types occurring in leaf litter, under loose bark, or hopping around on the surface often have long antennae and springs; some of these forms are brightly colored. Finally, there is quite a different kind of springtail occurring chiefly above ground, a type with a large head followed by a nearly spherical body bearing long legs, ventral tube, and spring. The Germans call these by the colorful name *Kugelspringer,* which might be translated "globe springer" or perhaps "jumping blob." Springtails of this type sometimes feed on seedling plants, and become a considerable nuisance. The so-called "Lucerne flea," a pest of some importance, is a member of this group.

Much has been written on the colors of springtails. The majority are whitish, slate-gray, or brown, but some are yellow, green, blue, lavender, or red; speckled, blotched, or banded; almost anything imaginable. Lubbock remarked that some of the brightly patterned globe springers resembled certain small spiders common in leaf mold, and he supposed that the springtails mimicked the spiders and thereby gained protection from natural enemies. It has been suggested that some of the springtails with patterns of strongly contrasting colors actually exhibit "warning coloration"; that is, several species have a common pattern easy for predators to recognize and remember. We know that many springtails are distasteful to predators (such as ants and certain mites). Some of them, in fact, exhibit what is called "reflex bleeding"; when attacked they exude some of their distasteful or even toxic blood through small pores in their skin. Most such species lack a well-developed spring, and seem to rely upon chemical defense rather than upon quick escape. A few springtails are luminous when disturbed. It is said that their entire bodies glow for five to ten seconds, although we know little regarding the origin or function of the light. Apparently it plays no role in courtship (as it does in fireflies), and it is conceivable that it may serve as a warning to nocturnal predators.

The colors of springtails may be produced by the skin itself or by an overlay of scales not unlike those of butterflies. Scales occur mostly on species living on top of the ground or in moderately dry soil, and it is believed that they may assist in retarding evaporation from their body surface. Scales may, however, play an equally important role in escape. Various small spiders spin webs close to the ground, and running and leaping springtails are apt to become entangled in them. Thomas Eisner, of Cornell University, has shown that the scales of butterflies and moths assist them in escaping from spiders' webs. Since the scales are only loosely attached, they tear off and stick to the adhesive strands of the web, coating the strands so that the insect can slip between them, minus a few expendable scales. Scales also render the body too slippery to be readily grasped by predators. Dr. Eisner watched silverfish being attacked by ants. In the few cases in which the ant was able to grasp the silverfish, the latter promptly slipped out of its grasp, leaving the ant with a "mouthful of scales."

We know that ants are major enemies of springtails. In fact, one group of hunting ants appears to specialize on them. These ants are equipped with enormous mandibles that can be opened to almost 180 degrees. As the ant stalks its prey, with its jaws agape, two long hairs extend from the front of the head. In the words of Edward O. Wilson, of Harvard University, who has vividly described this behavior, these hairs "apparently serve as tactile range finders for the mandibles. When they first touch the prey . . . a sudden and convulsive snap of the mandibles literally impales it on the teeth." The springtail, of course, tries to bring its spring into play—but the ant counteracts this by immediately lifting the springtail off the ground and paralyzing it with its sting. Springtails, in their turn, have experimented with a variety of escape and defense mechanisms of variable effectiveness. Evolutionary "cat and mouse" dramas such as this are by no means uncommon.

Just how important a role springtails play in the economy of nature is somewhat uncertain. Apparently they assist materially in breaking down fallen leaves and other detritus into humus. Some feed on the fungi that develop on plant detritus, while others subsist mainly on the fecal pellets of other soil animals. A few of the "globe springers," as already mentioned, are attracted to seedling plants, and others are pests in mushroom cellars. At least some springtails feed upon those ubiquitous creatures of the soil, the nematodes (or roundworms). William L. Brown, Jr., of Cornell University, watched a small colony of springtails under a microscope. He noted that "when a springtail came upon a nematode, even a worm nearly as long as itself, the worm was on each occasion seized at one end and rapidly and smoothly ingested, although still undulating rapidly until the free tip disappeared." Dr. Brown has told me that the springtails reminded him of children in an Italian restaurant, each gleefully sucking up his strands of spaghetti.

Certain African springtails appear to be restricted to the nests of termites of the genus Bellicositermes. They are said to ride around on the heads of the soldier termites. These soldiers have mandibles so specialized for colony defense that they cannot feed themselves. They are therefore fed by the worker termites, and when they are fed the springtails are right there poised over their mouths, ready to share in the meal.

But all this is rather exceptional. Most springtails live obscure and uneventful lives that may sometimes last as much as a year, the females from time to time laying a small cluster of eggs. The eggs are at first small and glistening, but they immediately absorb water and increase in size. Within a short while the outer shell bursts, and after that the developing springtail is protected only by the delicate inner membrane, which, surprisingly, bears large hairs in some species. After several days a small springtail looking like a miniature copy of its parents emerges. As it grows, the springtail casts off its skin periodically, and even after reaching sexual maturity it continues to molt, in contrast to nearly all true insects.

Springtails seem to make a fetish of cleanliness, a bit like cats and some humans. Perhaps this is necessary because most springtails have no special breathing tubes, and must absorb air directly through their delicate skins. Charles Macnamara described the manner of cleaning as follows:

"From its mouth it extrudes a small, bright drop of liquid, and taking it on the claw of one of the forelegs—where it looks like a gleaming boxing glove—it rubs it briskly over its antennae, head and legs. Sometimes it transfers the drop to the claw of one of the second pair of legs so as to reach farther down the body. The drop often remains unbroken during these proceedings, and with laudable economy, is then returned to the mouth and swallowed again."

The mating behavior of springtails has attracted much attention, particularly since neither sex has anything at all in the way of external genital organs, and the male therefore has no way of placing his sperm inside the female. Nevertheless, in some of the common globe springers, as Lubbock remarked, "the males are very attentive to the females, and caress them lovingly with their antennae." Further on he describes courtship in the following words:

"It is very amusing to see these little creatures coquetting together. The male, which is much smaller than the female, runs around her, and they butt one another, standing face to face, and moving backwards and forwards like two playful lambs. Then the female pretends to run away and the male runs after her, with a queer appearance of anger; gets in front and stands facing her again; then she turns coyly round, but he, quicker and more active, scuttles round too, and seems to whip her with his antennae; then for a bit they stand face to face, play with their antennae, and seem to be all in all to one another."

A good deal more is known about the reproductive behavior of springtails than was known in Lubbock's day, and the added knowledge does nothing to remove the air of improbability from their remarkable performances. One of the best general accounts is that of Friedrich Schaller, of the University of Erlangen-Nurnberg, Germany, in his little book *Die Unterwelt des Tierreiches* (The Underworld of the Animal Kingdom)—a

The courtship of Sminthurides aquaticus. The male (on the right) is smaller than the female and has a grasping organ on his antennae, making a prolonged tête-à-tête possible.

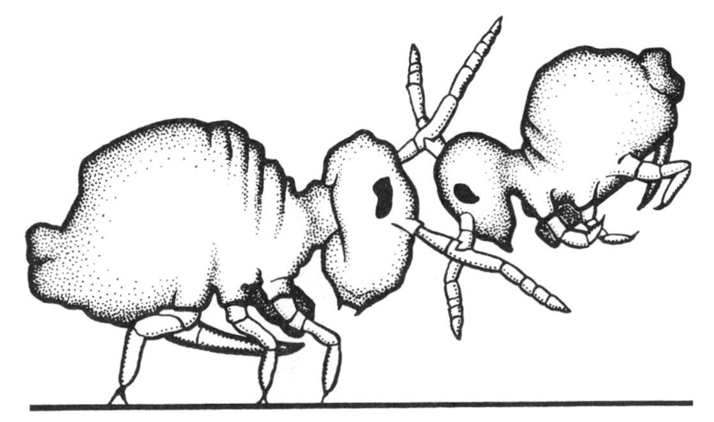

book that fortunately is now being translated into English. One of the better studied species is a small globe springer that frequents the still surfaces of small ponds and goes by the appropriate name Sminthurides aquaticus. The males of these tiny yellow springtails are much smaller than the females, and have curiously distorted antennae, two of the segments being opposable and furnished with a number of hooklike bristles. Using these peculiar grasping organs, the male takes hold of the base of the antennae of the female and is literally lifted off the ground by his much larger spouse. In Professor Schaller's words:

"While the female runs around apparently undisturbed and feeds on duckweed, the male hangs helplessly suspended in the air or lies facing backward against the head of his partner. Nevertheless in due time he takes command of the situation, letting himself down and beginning an often very laborious-looking 'dance' with his 'giant lady.' He goes forward and backward, to and fro. If she follows willingly enough, he deposits a small droplet of semen and pulls his partner towards it. This may happen in one of two ways: (1) the male simply goes backwards, leading the female by her antennae, or (2) he stays on one spot and swings around 180° with her. In either case the male seems to undergo great exertion. But most individuals reach their goal: at the right spot the female can contact the droplet of semen with her vulva. In the second case things sometimes go awry when the size difference between the partners is great. Then the fulcrum of the pair is so eccentric that the female ends up with her vulva beyond the sperm droplet and so cannot find it."

Only a very few male springtails have grasping organs on the antennae and are able to maneuver the female about in this manner. The males of another globe springer, Dicyrtomina minuta, which lacks these grasping organs, proceed in a different but perhaps even more remarkable way. This species occurs on tree trunks, where the females may often be seen feeding, resting, and cleaning themselves with no apparent concern for the much more restless and active males. From time to time a male approaches a female and touches her with his antennae. If she appears willing to sit reasonably still, the male then proceeds to place a whole ring of semen droplets around her; he is able to produce fifty or more at one time. Each droplet is placed on top of a short stem fastened to the substrate, a bit like dewdrops adhering to plant hairs. The female finds herself surrounded by a "picket fence" in which each post is surmounted with a blob of semen. The female's virginity is "at stake" in a real sense, and apparently few females maintain it for long under these conditions. It all seems like a perverted version of Brünnhilde surrounded by her ring of fire, although one somehow cannot imagine the male springtail hopping about to Siegfried's *leitmotif.*

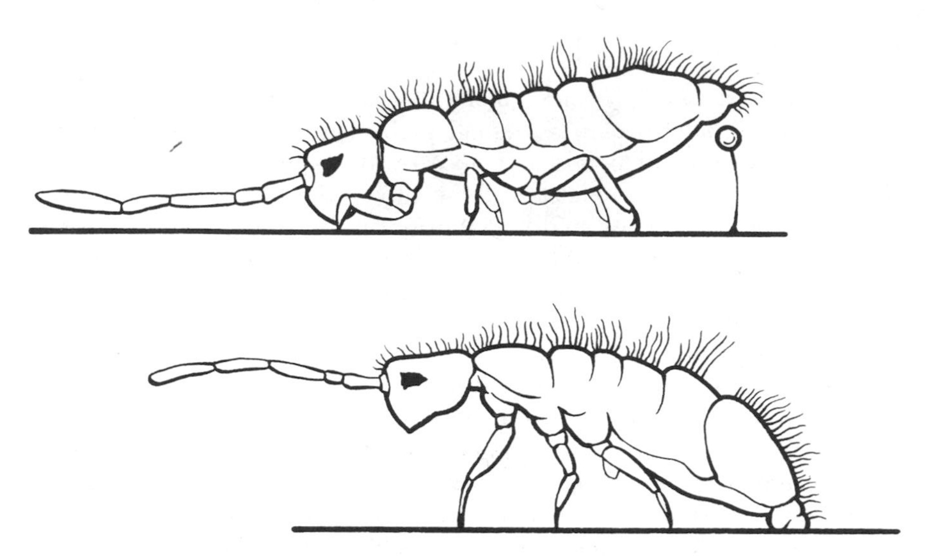

Mating by "remote control" in a springtail. At the top, a male deposits a droplet of semen attached to a stalk. At the bottom, a female blunders into the droplet and gathers it into her vulva. (After Friedrich Schaller.)

Professor Schaller speaks of these as cases of "half-pair" formation, since the female's role is wholly passive, while the male works very hard to ensure her fertilization. In the majority of more ordinary springtails there is scarcely anything that could be called pair formation at all, for the male scatters about his stalked droplets of semen in the total absence of females, but of course in places where females are likely to blunder into them. The males of some species plaster the earth with so many of these stalked droplets that they resemble a growth of mold. The female now strolls in the "love garden" planted by the male, and the transverse slit that forms her vulva sooner or later picks up one of the many valentines awaiting her. Actually, it is probable that any bodily contact causes the droplets to burst. The sperms may then travel to the vulva under their own power, or perhaps the female aids them with cleaning movements of her ventral tube.

Various persons have noted that the males often browse in their gardens and actually eat some of these droplets of semen. Helmut Mayer, of the University of Mainz, Germany, has found that each time one droplet is eaten, another is deposited. He has found that in fact the sperm in these droplets remain potent for only about eight hours. The male apparently goes about and devours those droplets which are more than eight hours old and replaces each with a fresh one. Thus the female is always confronted with viable sperms.

Incidentally, some of this behavior was observed long ago but misinterpreted. For example, Victor Lemoine reported to the French Association for the Advancement of Science in 1883 that the male takes the sperms into his mouth and then applies his mouth directly to the vulva of the female. Many persons have tried to repeat Lemoine's observations without success, and it now seems probable that he was simply incorrect in some of his details. Study of the sexual behavior of springtails has always been handicapped by the fact that males and females of most species look exactly alike. Indirect transfer of sperms can, however, be demonstrated if one isolates individual springtails in small rearing chambers. Those that produce semen droplets are, of course, adult males. Having obtained some semen droplets, one may then remove the male and shortly thereafter introduce an individual that up to this time has produced no eggs or sperm. Like as not she will begin laying fertile eggs within a few days, indicating that she has been fertilized by "remote control."

Indirect sperm transfer poses some interesting problems. For instance, we know that rich soil harbors many species of Collembola. In his studies of the springtails of the forest floor in the Cathedral Pines, a group of virgin white pines in Connecticut, Peter Bellinger collected about sixty species. If many of these species place their semen droplets at random on the substrate, how do the females of each species pick up the sperm of their own species? What prevents the formation of hybrids? The differences in courtship and insemination that we think of as being characteristic of sexually reproducing animals are completely absent; we don't even know if the males recognize females of their own species. Presumably there are genetic mechanisms at work: the "wrong" sperm simply don't bring about proper development of the egg.

Another problem is posed by the fact that indirect sperm transfer is characteristic not only of springtails but also of many quite unrelated groups of animals occurring in the soil: for example, bristletails, some centipedes and millipedes, and many mites. Some of these groups produce stalked semen droplets strikingly similar to those of springtails. Is this simply a result of the fact that this type of reproduction is most efficient in the confined habitat of the soil? Or is it rather a reflection of the fact that all these are the little-changed descendants of very primitive groups of animals? Friedrich Schaller points out that the forefathers of all earth animals came from the water, and in water fertilization of the eggs outside the body of the mother is very common, although the sex partners must be close together to avoid great dilution of their products. When these primitive arthropods came out on land, perhaps it became feasible for the males simply to drop their sperms when the females were not even there, so long as they were scattered about in abundance in places where females would encounter them. The course of evolution in the Collembola is apparently in the direction of more direct contact between the sexes, for the globe springers are among the most highly specialized springtails in every respect. This lends support to the belief that these various types of soil animals are all hanging on to indirect sperm transfer as an inheritance from their very early common ancestor. Of course, it *does* work: after all, we are talking about the most abundant of all land animals! And doubtless the development of very similar stalked semen droplets in unrelated groups is a matter of convergence: that is, this particular "gimmick" proved effective in soil habitats, and was arrived at by several groups independently. So both of the answers we suggested earlier are probably in some measure correct.

Speaking again of the abundance of springtails leads us to the final problem worth thinking about: Why do springtails sometimes swarm in such incredible numbers that even the casual observer is suddenly aware of these tiny animals? Swarming is a common occurrence, and has been reported many, many times, in such places as the margins of ponds, tree trunks, manure or rich leaf mold, or lawns and gardens. But it is most often noted when these normally secretive animals come piling out on the surface of snow by the millions. These are the so-called "snow fleas," so well known to those persons who, like myself, take advantage of warm days in January or February to restore the circulation in their limbs (and brains). On a January day in 1860, Thoreau made the following entry in his journal:

"The snow flea seems to be a creature whose summer and prime of life is a thaw in the winter. It seems not merely to enjoy this interval like other animals, but then chiefly to exist. It is the creature of the thaw. Moist snow is its element. That thaw which merely excites the cock to sound his clarion, as it were, calls to life the snow flea."

Charles Macnamara, a springtail enthusiast whom we have met before, writes of the vast swarms that "literally blacken square yards of the snow around the principle foci from which they emerge. On the level surfaces they may be as thick as 500 to the square foot, while in hollows and depressions in the snow—such as foot-prints—from which they cannot easily escape, they sometimes accumulate in solid masses that could be ladled out with a spoon." As Macnamara and others have pointed out, the snow flea is not a particular species of springtail, but any of a number of species that occasionally swarm on the surface of snow. Most of these are dark in color, but in western Canada there is a species, called the "golden snow flea," that is said to cover the snow with a carpet of gold.

Macnamara noted that our commonest snow flea, Hypogastrura nivicola, has an odor resembling that of sliced raw turnips, and he was sometimes able to locate swarms by tracing whiffs of turnips across the winter snows. Macnamara found that snow fleas come up through the snow by following the small spaces around the stems of weeds and shrubs. When bad weather once again forces them down, those that have remained close to plants go back the same way, but many others attempt to work their way straight down through the snow. The vast majority of these perish when they strike layers of compacted snow and cannot reach the earth. Macnamara noticed that he never found a swarm of snow fleas in the same place twice, suggesting that "a great destruction of the insects must take place on every occasion."

Springtails have been reported to form fantastically populous swarms elsewhere than on snow. The notorious "Lucerne flea" sometimes occurs in clover and alfalfa fields at a density of six thousand per square foot. D. L. Wray, one of the current American specialists on springtails, once reported springtails in fields and tobacco seedbeds in North Carolina "so thick as to form blue blankets of insects inches thick." In his book *Animal Ecology*, the British biologist Charles Elton cites examples of population explosions, including the following:

"In Switzerland the railway trains are said to have been held up on one occasion by swarms of Collembola or springtails, which lay so thickly on the lines as to cause the wheels of the engines to slip round ineffectually on the rails."

What causes these outbreaks? Presumably they are preceded by a period in which food, temperature, and moisture are just right in a given situation, permitting a rapid increase in numbers, at the same time causing a decline or at least no increase in the numbers of their natural enemies. We know that springtails reproduce very rapidly under ideal conditions. One group of workers started with a laboratory colony of 10, and in 8½ weeks found that it had increased to 3,447. Such a build-up in nature may well result in exhaustion of the local food supply. Under these conditions there may be behavioral changes that result in mass migrations. Studies of migrating masses show that, rather than shunning light as springtails usually do, they are attracted to brighter light intensities. Once out of their original habitat, they may perish in great numbers, as described by Macnamara for snow fleas. W. M. Davies, in England, studied a swarm under a microscope, and found great numbers of partially consumed bodies:

"Individuals could be seen vigorously attacking other members of the swarm. . . . During an attack the victim struggled and fought with mouthparts and tarsal claws, but was gradually overpowered. When it succumbed, several other members of the swarm quickly collected around its mutilated body and devoured it. . . . Individuals about to moult proved ready prey for the more active members of the swarm."

Recent studies suggest that more commonly the springtails merely feed upon the bodies of those already dead—but since the blood is toxic, they, too, are killed, and the result is a massive population crash. Perhaps the swarms of springtails are not very different from the well-known irruptions of lemmings; that is, they appear to "go berserk" under conditions of high population pressure. Evidently they enter situations where they perish in great numbers, or perhaps they often simply eat each other, either alive or dead. This seems a crude method of population control, but one assumes that some of them do find new sources of food. Even humans have been known to resort to cannibalism when all their food is gone. If human populations reach the point of exhausting the resources of the earth because of religious and social taboos on control of family size, it would be interesting to see to what extent the last survivors uphold the taboos on cannibalism and the eating of human corpses.

Of course, we do not really know the cause of these outbreaks, and we do not really know if mass destruction and cannibalism are inevitable results of swarming: we know only that swarms appear and disappear and that cannibalism and mass death do sometimes occur. We know pitifully little about population phenomena in animals generally, and we hardly know the basic facts of life about most of these incredibly abundant soil animals. Springtails occur all over the earth, from deep in the soil to high in trees, and specimens have even been recovered from samples netted from airplanes two miles high in the air! They are common inhabitants of caves, and one "ghastly white" species, Isotoma sepulcralis, "makes its hideous habitation with moulding human bodies in the grave" (Macnamara's words). Some occur on the surface film of ponds, others on the wrack of sea beaches, where they survive immersion in salt water for considerable periods. The "glacier flea" lives on glaciers in the Alps and subsists on windrows of pollen on their surface. Certain species inhabit the nests of birds and mammals, others the nests of ants or termites. Some of the agricultural pests are highly resistant to DDT and related insecticides, and in areas that are sprayed regularly or in which soil fumigation is practiced, the springtail population tends to increase, sometimes dramatically.

And yet we know so little about them! While we support vast programs designed to find out if there is a hardy microbe in some protected canyon on Mars, a scattered handful of underpaid and underequipped researchers struggle to decrease vast areas of ignorance about creatures that abound in our lawns, gardens, and woodlands. I am not suggesting that we junk our telescopes and abandon our space probes. We can afford both the sky and the soil, intellectually as well as financially. Right now, while we are in the midst of a population and technological explosion that is transforming the earth in ways we cannot predict, it would seem more urgent to study those little-understood animals on lower levels of the pyramid of life that supports us. They are not only marvelous creatures in their own right; they may also teach us something about population control and the proper utilization of the earth's resources.

The Intellectual and Emotional World of the Cockroach

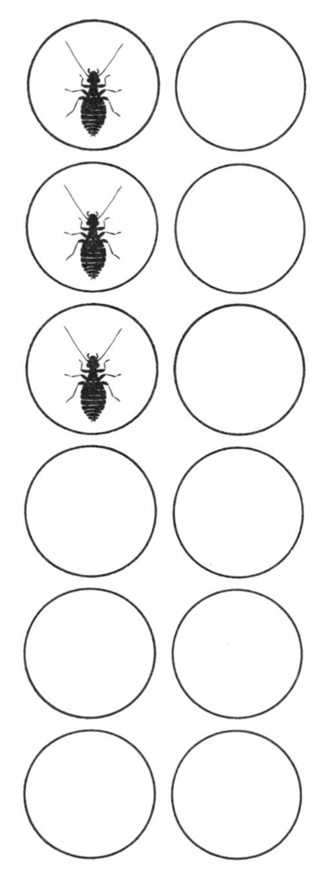

Ever since archy stopped jumping on the keys of Don Marquis' typewriter in the offices of the New York *Sun,* cockroaches have passed from the ken of most of us. It is a pity. Archy was not only a character in his own right; he also knew a wonderful assortment of fleas, spiders, rats, alley cats, and other reputedly despicable creatures. Ours is a world of insecticides, rodenticides, herbicides, and etceticides. As archy complained:

> i have just been reading
> an advertisement of a certain
> roach exterminator
> the human race little knows
> all the sadness it
> causes in the insect world . . .

Of course, a biologist will tell you that insects are unlikely to experience sadness and that archy was not a *real* cockroach. Perhaps so; but there is a sadness here, for the human species is bereaved when it is unable to appreciate the world of small and creeping things. I heartily recommend cockroaches. Unlike archy, the average roach has little or no poetry in his soul. But he is a marvelous beast nonetheless. He must, of course, be met on his own terms, in his own world. He has been inhabiting that world successfully for somewhat more than 250 million years. The earliest fossil cockroaches look so much like contemporary species that one can almost imagine them freshly crushed by some irate housewife. But the first housewife was still more than 249 million years in the future. Any creature so adept at survival would seem to be worth our attention; survival is a subject we can afford to learn much more about.

The cockroach is the only animal in which both male and female are rightfully called "cocks," since the word is a corruption of the Spanish word *cucaracha,* and the word "roach" merely a further corruption of "cockroach" (a lesson in etymological entomology; or is it the other way around?). A roach, properly speaking, is a kind of fish, and an author who writes an article on roaches usually finds it filed under "fish" in the library. Having used the word "cockroach" in the title, I shall often say "roach" for short. Some people call them "water bugs," for no good reason at all, so why not roach?

Cockroaches are primarily creatures of the tropics and subtropics, and in temperate regions we know them mainly from a few species that have found an easy living in our homes, stores, and restaurants. Domestic species include, among others, the American, German, Oriental, Surinam, and Cuban cockroach. (A house my family once rented in Florida was a veritable United Nations of roachdom.) I once had a cockroach served to me in an order of beefsteak and onions in Texas. (I believe it was an American roach, but accurate identification of fried specimens is difficult.) I was ravenously hungry after a day in the desert, so I cleaned my plate except for the cockroach, which I spread out neatly in the center of the empty plate, arranging his antennae and legs as best I could. The expression on the waiter's face when he picked up my plate was ample compensation for the health risk I took. Cockroaches are basically clean animals, but they do track about a good deal of human filth; domestic roaches have in fact been found to carry bacteria responsible for a variety of intestinal disorders, as well as polio virus and even hookworm. I survived my Texas meal well, although my sorting out of fried onions from fried roach parts was sometimes arbitrary.

Such experiences are rare nowadays, but it was not always so. After a trip to western Canada in 1903, the entomologist A. N. Caudell wrote to the journal *Entomological News* as follows:

"Cockroaches thrive in British Columbia, as they do almost everywhere. The common species seems to be the German roach. . . . They are in everything, even the food. On this trip I had them served to me in three different styles, alive in strawberries, à la carte with fried fish, and baked in biscuit."

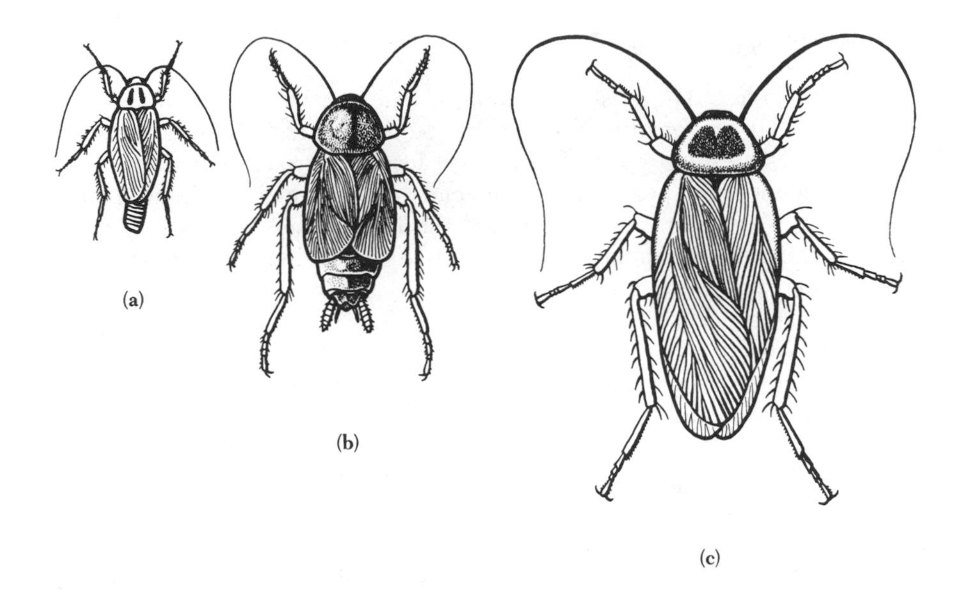

Three prominent members of "the United Nations of Roachdom," drawn to the same scale: The German roach (*a*), the Oriental roach (*b*), and the American roach (*c*). Of the three, the German roach is the only one to carry about her egg case until hatching time.

The names of our domestic roaches are largely the result of chance. When the Swedish naturalist Linnaeus received a roach from America, he called it "americana," while a roach from Asia he called "orientalis." Even by that time (1758) most domestic cockroaches had been spread over much of the globe, and modern transportation has finished the job. The late James A. G. Rehn, of the Academy of Natural Sciences of Philadelphia, was, among other things, something of a Sherlock Holmes of man's cockroach camp followers. His studies revealed that the American roach and its close relative the Australian roach belong to a group that occurs in the wild primarily in tropical Africa. He felt that these species, along with the Madeira roach and several others, came from Africa to America at an early date on slave ships. The Oriental roach also has wild relatives in Africa, but the evidence suggests that it arrived in Europe very long ago indeed, perhaps on Phoenician vessels. Later it was apparently carried to South America on Spanish galleons and to North America on English ships.

The German roach, according to Rehn, came from North Africa, but at a somewhat later date than the Oriental roach. During its spread across Europe it was called the "Prussian roach" by the Russians, the "Russian roach" by the Prussians, thus paralleling the history of syphilis, which was known as the "French disease" throughout much of Europe, but the "Italian disease" by the French. The first outbreak of syphilis in the British colonies in America, by the way, occurred in Boston twenty-six years after the landing of the *Mayflower*. Evidently that noble ship and its immediate followers carried a good many things besides bluebloods, including, no doubt, the German roach, long an inhabitant of Boston slums but now fighting a rearguard action against urban renewal and the more recently arrived brown-banded cockroach. The German roach is now one of the most ubiquitous of man's unwanted pets.

America does, of course, have native roaches, but few of these have become domesticated, perhaps a reflection of the fact that man himself had his origins in Africa, thus giving the African roaches a big head start. The so-called Surinam roach apparently did not come originally from that Dutch colony in South America; Rehn found its closest relatives in the Orient, whence the species apparently spread into Africa, and then joined several other species in slave ships traveling to that brave new world America. Only one species, the so-called pale-bordered cockroach, has reversed the usual direction of immigration (may I say encroachment?) and reached the Canary Islands from its home in the West Indies.

Biologists are always on the lookout for animals easy to rear in the laboratory, and what could be easier than cockroaches, which are usually there to start with anyway. Some laboratories now have cultures of dozens of species of cockroaches, not all of them domestic species. Most of them require no more than a warm and cozy cage, a little water, and an occasional dog biscuit. Best of all, cockroaches are exempt from P.L. 89-544, the Laboratory Animal Welfare Act, passed by Congress in 1966 to regulate the handling of "dogs, cats, and certain other animals" used in research and experimentation. Evidently the cockroach lovers have no lobby in Washington.

Cockroaches have played a role in many basic studies of animal behavior, nutrition, and metabolism. They have even been found useful in cancer research. Berta Scharrer, of the Albert Einstein College of Medicine, found that when she cut certain nerves in the Madeira roach they developed tumors in some of the organs supplied by those nerves. Other workers have found tumors resulting from hormonal inbalance after transplanting endocrine glands in roaches. The application of these findings to the understanding of cancer in humans remains to be seen.

Behavior studies suggest that roaches are among the "brighter" insects. This was demonstrated, in 1912, by C. H. Turner, of Sumner Teachers College in St. Louis, a man whose ingenious studies of animal behavior, often conducted with homemade equipment, earned him a reputation as one of the leading Negro biologists of his time. Turner, for example, tried "teaching machines" on cockroaches long before they came to be used for humans. He put roaches in cages containing two compartments, one lighted and one dark. True to their well-known preference, Turner's roaches regularly headed for the dark compartment. However, when he wired this compartment in such a way that they received an electric shock upon entering, they quickly learned to go directly into the lighted compartment. The males, he found, seemed to learn more rapidly than the females.

Recently some of the potent new insecticides have been found relatively ineffective against cockroaches. Actually, these insecticides kill roaches quickly when the insects are forced to remain on sprayed surfaces. But in actual use, the chemicals have a repellent effect, and the roaches quickly learn to avoid cracks and crevices that have been sprayed and to fan out into other, unsprayed areas. A modern equivalent of Turner's "teaching machine" showed that roaches readily learned to associate darkness with the presence of insecticide and to choose the lighted, unsprayed compartment.

Cockroaches are among the few insects that can be taught to run mazes successfully, as Turner was also one of the first to find out. He rigged up a complex pattern of pathways made of copper strips supported over a pan of water. At the end of one runway was an inclined plane leading to the jelly glass that was "home" to that particular roach. After only five or six trials at half-hour intervals, most roaches greatly reduced the time required to reach their jar as well as the number of errors committed. In the course of a day the number of errors declined to almost zero. Turner's Oriental roaches had short memories and had to be retrained every day, but another worker found that American roaches showed more day-to-day retention and even improvement. Other researchers tried running two or three roaches at a time to see if they could solve a maze more rapidly in company, as certain fishes can. Exactly the opposite occurred. Apparently extracurricular activities sometimes conflict with serious training even among roaches.

Lest anyone be inclined to dub the roaches "eggheads," I hasten to add that roaches *without* their heads are able to learn some things well. Recently G. A. Horridge, of St. Andrews University in Scotland, arranged a decapitated roach in such a way that the legs received electric shocks every time they fell below a certain level. After about thirty minutes the roach changed its behavior in such a way that the legs were raised and few shocks received. A decapitated roach, by the way, often lives for several days, although it eventually starves to death.

Doubtless the learning abilities of roaches have something to do with their success in putting up with the shenanigans of mankind. Other reasons for their success are to be found in their ability to scuttle off rapidly into crevices and to remain remarkably alert to peril. The roach's alarm system consists of very long and active antennae on his head, and a pair of similar but shorter structures at the other end of the body, called "cerci" (the Greek word for tails, of course much more erudite than calling them "tails"). These cerci are remarkably sensitive structures, and a light puff of air directed at one is sufficient to send the roach scurrying. Close study shows that the cerci are covered with tiny hairs that bend when a current of air strikes them. Deflection of the hairs stimulates some of the many nerves in the cerci, which send a message to two clusters of nerve cells at their base. Here the message is transferred not to ordinary nerves but to "giant fibers" many times larger than ordinary nerve fibers. The abdominal nerve cord of the cockroach contains several such giant fibers, each of them much larger in diameter than even the nerve fibers in our own bodies. These giant

Roeder's apparatus for testing the speed at which the American cockroach responds to a stimulus applied to the cerci. A jet of air (*j*) is discharged so as to strike the cerci of the roach and at the same time move a paper flag (*f*) connected to an oscilloscope (*o*) via an amplifier (*a*). The roach is suspended from a wire also connected to the oscilloscope, and allowed to grasp an unattached ball of cork (*c*). (Adapted from Kenneth D. Roeder, *Nerve Cells and Insect Behavior* [Harvard University Press, 1963]; used by permission of the President and Fellows of Harvard College.)

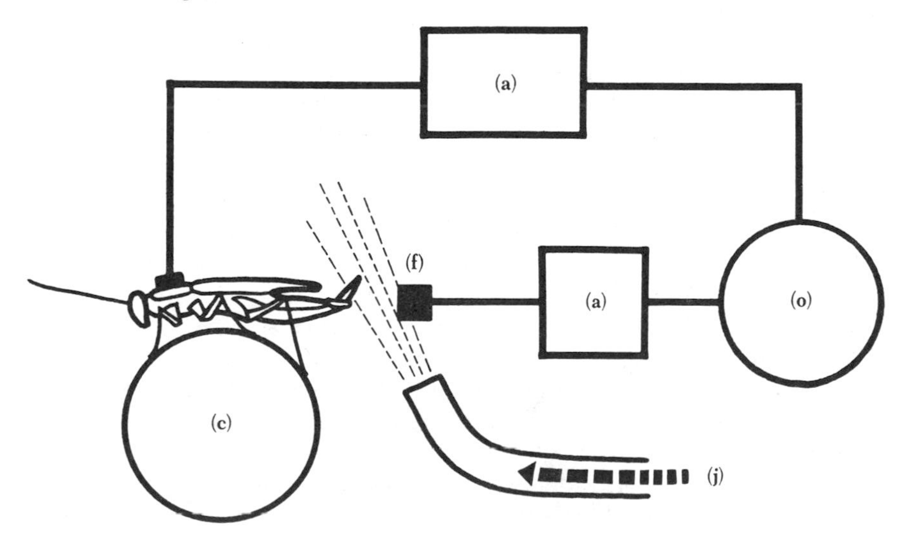

fibers carry nerve impulses more than ten times as fast as ordinary nerves, actually at a rate of more than fifteen feet per second (less than .003 seconds over the length of a giant fiber of the American roach). These fibers carry the impulse directly to the nerves and muscles of the legs, and produce the apparent immediate escape response so characteristic of roaches.

Kenneth Roeder, of Tufts University, has made a detailed study of evasive behavior in the cockroach. In the course of this study he rigged up a treadmill attached to a very sensitive recording device. The roach was placed on the treadmill, and just behind it was placed a small tube through which a jet of air could be blown at the cerci. At the same time the air jet would strike a small paper flag also connected to the recording device, such that the interval between air jet and leg movement was registered. The cockroaches were fairly uncooperative, as experimental animals often are, and frequently proceeded to clean themselves or make other unscheduled movements. But eventually Professor Roeder obtained twenty-three good measurements that averaged out to about .05 seconds from air puff to leg movements. In subsequent experiments he found out why, although transmission over the giant fibers requires only about .003 seconds, another .047 seconds, more or less, are required for the final response. Some of the difference, he found, was made up by "synaptic delays," that is, the time taken for the impulse to cross from one nerve to another.

Synapses are the switchboards of the nervous system, and provide the major means of sorting and directing messages. They do slow things down; but giant fibers by-pass many synapses and provide for a much quicker response than if a series of smaller nerves were involved. As Roeder points out in his book *Nerve Cells and Insect Behavior*, the number of ordinary nerve fibers that would occupy the space of a giant fiber might handle a much greater amount of information, but at the cost of several thousandths of a second. In the course of evolution, this small gain in speed of escape from enemies outweighed the importance of carrying more detailed messages.

Our own human warning systems operate on much the same principle: emphasis is on rapid transmission of simple messages ("missile approaching") rather than much slower transmission of analytical reports. Such a system may have enhanced the survival of roaches as a group by millions of years, for their response is quick escape, and if the source of stimulation is in fact harmless, nothing is lost. *Our* problem, since we have no place to escape to, is to avoid an inappropriate response to meaningless information.

Not only are roaches adept at quick escape, but some species have developed mechanisms for spraying would-be predators with repellent chemicals. The giant Florida roach and several other species have large glands from which they eject to a distance of an inch or two a foul-smelling substance similar to that produced by stinkbugs. The roach Diploptera punctata, however, has glands of an entirely different nature and producing very different chemicals, quinones similar to those of bombardier beetles and certain millipedes. Thomas Eisner, of Cornell University, who has made intensive studies of the defense mechanisms of insects, found that the spray of this roach caused attacking ants and beetles to retreat and to undergo "a series of abnormal seizures, during which leg movements became discoordinated and ineffectual." Incidentally, quinones similar to those produced by certain cockroaches are known to have bactericidal properties, and may someday conceivably find a role as medical antibiotics.

At least one roach has wholly abandoned cowardice in favor of aggression. This roach, with the suitably frightening name Gromphadorhina portentosa, not only produces an odor but also makes a loud hissing sound when disturbed. The males, which sometimes reach a length of four inches, are unique in having a pair of thick horns just behind the head. When males chance upon one another they charge and push each other back and forth with their horns, all the while hissing loudly. This roach is a native of Madagascar and has not become domesticated, thank God: it is not the sort of thing one would want to encounter on one's kitchen shelf.

It is in their sex lives that cockroaches best reveal the secret of their success in keeping the world populated with their kind. Like other insects, roaches have no hormones produced by their sex organs. They hardly need them, adult insects being designed for reproduction and not much else. They do have certain built-in inhibiting devices, however; insects cannot afford to spend *all* their waking hours in sex, phonetics notwithstanding. We know that certain endocrine glands in the head of a female roach have much influence on the formation of her eggs. Also, we know that, in some roaches, if these glands are removed soon after the female becomes sexually mature, she fails to produce a chemical necessary to attract males, and is therefore very likely doomed to spinsterhood. However, the sexual behavior of these females is otherwise normal; it is possible to douse them with sex attractant taken from normal females, and they then attract males and mate in the usual manner. When a female Surinam roach is pregnant, pressure of the developing eggs sends a nervous impulse to the head, suppressing these same glands and thereby stopping production of the sex attractant until the eggs are laid.

Many insects besides roaches are known to produce chemical substances that influence the behavior of other members of their species.

The study of the chemistry and effects of these "pheromones," as they are called, is one of the very active fields of biology at the present time. In some cockroaches, the male must actually contact the female before being stimulated by substances on her body, while in other cases the female produces a pheromone that attracts males from a distance. The parallel with females of the human species, who unfortunately must rely upon the Revlon Company and its competitors for their pheromones, is obvious.

The German roach has paid for its intimacy with man by having its sex life analyzed in infinite detail. When a male detects the pheromone of a female, he faces his intended spouse, and the pair begin to "fence" with their antennae. Shortly thereafter he turns completely around and faces away from her, at the same time raising his wings at about a ninety-degree angle. This may seem an impolite gesture to a human observer, but in fact the male is offering the female his own particular chemical attractants, which exude from glands on his back. If courtship is proceeding well, the female climbs upon his back and begins to feed on these exudates, which apparently function to lure the female into copulating position. After a few seconds the male begins to push himself farther back beneath the female, at the same time extruding his genital organs. These are extraordinary structures, resembling nothing so much as a Boy Scout jackknife, with its various blades and bottle and can openers. A successful male German roach manages to clamp on to a small crescent-shaped plate at the tail end of the female with one of the longest of these hooks, whereupon he moves out from under her and turns about facing away from her. At this time other, smaller hooks become attached to other structures on the female, so that she is "hooked" very literally for the hour or two required for copulation.

Louis Roth, of the United States Army Laboratories in Natick, Massachusetts, has been surveying the reproductive activities of various cockroaches, and has come to be called the Alfred Kinsey of the roach world. Fortunately, the roach's response to a questionnaire is to eat it. I say "fortunately," because it is much more fun (and more scientific) to study these matters directly. Some species undergo certain body movements in addition to exchanging pheromones. The male Madeira roach does a series of "push-ups" in front of the female, while the male Blaberus craniifer, one of the handsomest of roaches, is said to shake his abdomen and to butt the female with his head. The male American roach (somewhat embarrassingly) dispenses with all preliminaries and simply rushes at the female with his wings spread, "pushes his abdomen under the body of the female, and attempts to grasp her genitalia." In this species, as in many others, males are greatly stimulated by female sex pheromone even in the absence of females. If filter paper is taken from the bottom of

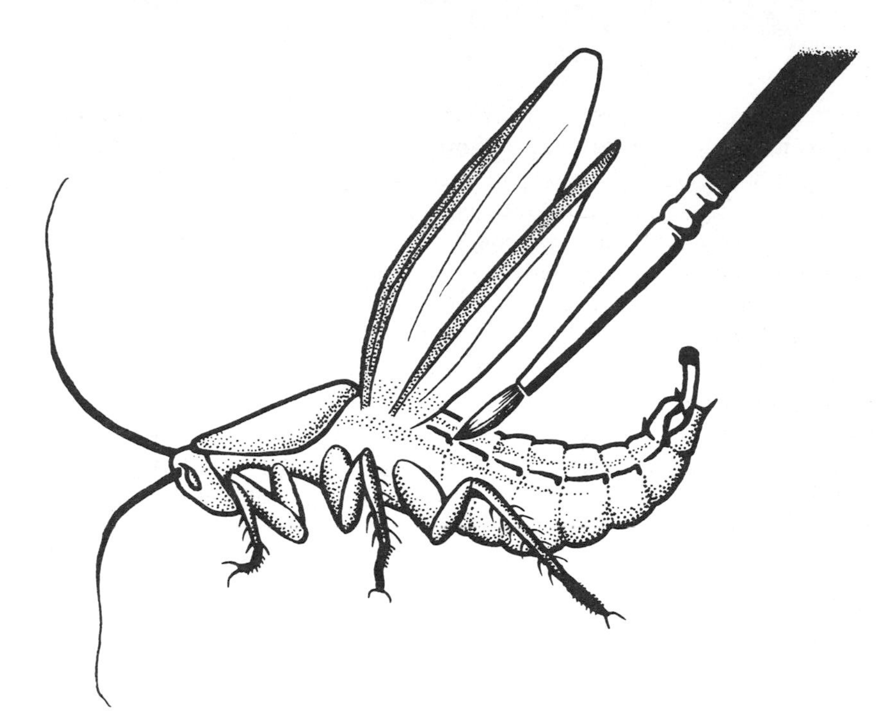

Artificially induced copulatory movements in a male cockroach. The female normally mounts the male and feeds at the products of glands on his back; this feeding induces the male to extrude his genital organs and grasp those of the female. This male has been placed in the presence of female hormone and he has raised his wings in response. When the experimenter strokes his back with a brush, the roach responds by thrusting his copulatory organs upward and backward. (After Robert Barth.)

a cage of females and placed in a cage of males, the latter become greatly excited, flutter their wings, and attempt to mate with the paper. Workers at the United States Department of Agriculture have succeeded in isolating the sex attractant of the American roach by passing a stream of air through jars containing thousands of females, collecting the vapor by freezing it with dry ice. In this way they obtained, in nine months, 12.2 milligrams (about .0004 ounces) of a substance that proved intensely exciting to males.

Robert Barth, of the University of Texas, has recently been probing still more deeply into these matters, and might be called the Sigmund Freud of the roach world. He has found a low incidence of homosexuality in several roach species. But when female pheromone is introduced into a cage of male Cuban roaches, they tear about their cage and proceed to court one another furiously. All of the steps seen in heterosexual courtship can be observed, except of course for the final hooking together of the genitalia. We still know very little about the chemical nature of these potent sex attractants. Perhaps it is just as well.

There is apparently at least one kind of cockroach in which the female does not produce a pheromone. A curious species that has recently attracted the combined talents of Drs. Roth and Barth is Nauphoeta cinerea. The male of this species courts any individual of either sex that comes along, touching it with his antennae and turning about and raising his wings so as to expose his back. A receptive female feeds on the surface of his abdomen, and the male then slides backward beneath her, in the manner of the German roach. Receptivity on the part of the female is believed to be controlled by a hormone secreted by cells in the brain. When a female has mated, she retains the sperm packet in her vulva for some time, and it is apparently the pressure of this sperm packet that sends a message to her brain and "shuts off" the hormone. Roth found that if he cut the nerve cord, females would show "hypersexual" behavior, which means simply that they would mate over and over again in rapid succession. We may feel fortunate that roaches are not more educable than they are. Were the males able to read and understand Roth's papers, they would surely try to emulate his techniques.

Roth and his colleagues have had some success in extracting the substance on the abdomen of the male Nauphoeta cinerea that attracts the female. When this substance (which they have called "seducin") is placed on a dummy in a cage full of females, the latter rush over to the dummy and commence to lick it. Curiously, extracts of the male pheromone of American and Oriental roaches and several other species all produced a response in female Nauphoeta cinerea. This is one of the few pheromones produced by male insects to have been studied, and it has raised hopes that the use of these substances may someday help us to lure and destroy female roaches. The use of female sex pheromones has the obvious drawback that only males are attracted; and as everyone knows it is females that are the root of all evil.

The male Nauphoeta cinerea has other tricks besides seducin in his repertory. If the female fails to respond, he stands back an inch or so and begins to tremble. This trembling causes a series of fine ridges on the back of the thorax to rub against another series of ridges on the base of the wings. The result is a series of chirps that may serve to rouse the female. Similar sounds are produced by both males and females of this species when they are handled, and are usually termed "protest cries." Apparently the male "doth protest" when the female is indifferent to his pheromone. The sounds he makes are by no means the orderly, musical notes produced by his distant relatives, the crickets and grasshoppers (discussed in Chapter 5). But evidently they do the trick.

The eggs of roaches are produced in neat little packets that, in our homes and laboratory cages, are simply dropped on the floor, to hatch some time later if they do not dry up or become food for another roach. But we now know that simple dropping of the egg capsules is, in most cases, abnormal behavior produced by an unnatural environment; in nature, most roaches conceal their eggs in one way or another. Light on this particular subject came to me from a Coleman lantern placed on the ground near our camp in a state park in Florida one warm spring evening several years ago. Around me were several female giant Florida roaches—a brown, wingless, and rather odorous species that has not become domesticated to any extent. Each had an egg case protruding from the end of her body, and each was digging a hole in the sand or at least looking for a place to dig. When a suitable place had been selected, each roach made a series of backward strokes with her head, piling the sand beneath and behind her. After a hole about a third of an inch deep had been completed, she changed her tactics completely, dribbling saliva into the hole and picking up the moistened sand grains with her mouth, eventually molding a trough-shaped cavity of proper size and shape to fit the egg capsule. Next she straddled the pit, released the egg case, and slid it into the hole with movements of her abdomen, turning around and making final adjustments with her mandibles. Then she picked up sand from the margins of the hole, moistened it with saliva, and plastered it over the top of the hole by picking up sand in one place and dropping it in another. Then, after more than an hour of hard work, she wandered off into the darkness, having effectively protected against predators, parasites, and desiccation offspring she would never herself see.

In laboratory cages, the giant Florida roach, like other species, merely drops her egg cases on the floor. But when provided with sand, she will act out her normal egg-burying behavior (to a bleary-eyed audience, some time in the middle of the night). But this behavior proves to be somewhat adaptable, and can be modified in such a way that effective concealment occurs when the roach is placed on a substrate of rotten wood. Frances McKittrick, then at Cornell University, studied similar behavior in other roaches, and found burying of the egg case to be widespread but to differ in details in various species. The most primitive roaches (that is, those that we believe to show the fewest changes since ancient times) drop their egg case before making a hole for it, dig mostly with their heads rather than their legs, and cover the egg case imperfectly. The giant Florida roach is moderately advanced, but it is surpassed by many species.

The most advanced roaches have, in fact, evolved some quite different methods of protecting their eggs. The German roach carries her egg case around, projecting from the end of her body, and even transfers water to it, finally dropping it when the eggs are ready to hatch. The German roach and its relatives rotate the egg case ninety degrees before extruding it; that is, its greater diameter becomes parallel to the ground. This has preconditioned the higher roaches for a still more elaborate method of egg protection: the egg case, after extrusion, is drawn back into the body, where it occupies a special brood sac until hatching occurs and living young emerge from the mother. Without this prior rotation, the egg case would scarcely fit into the flattened body of the roach. As Louis Roth has pointed out, these roaches are unique in being "born twice," since the eggs first leave the body of the female and are then drawn back in, to emerge a second time as young roaches. In the brood sac, the eggs are thoroughly protected and are supplied with water and even, in at least one case, with nutriment. Roaches that give birth to living young include the Madeira roach, the Surinam roach, and several others. The Surinam roach has even dispensed with the nuisance of having a male sex; one strain of this species consists entirely of females that produce live female young, which grow up to produce live female young, and so on ad infinitum. If there is a more efficient reproductive mechanism, the roaches will undoubtedly find it.

Our affluent society may have banished the roach from its thoughts, but the roach is still a very long way from being ready to turn up his tails (his cerci, that is). Louis Roth and Edwin Willis concluded a recent survey of the relation of roaches and men with the comment that "cockroaches are tough, resilient insects with amazing endurance and the ability to recover rapidly from almost complete extermination. They will probably always be with us, and we can only temporarily reduce their numbers." They describe a four-room apartment in Austin, Texas, that in 1947 was found to contain between 50,000 and 100,000 roaches, mostly German. The apartment was sprayed, but six months later at least 15,000 German roaches had reestablished themselves. They also present a striking photograph taken quite recently of the walls and roof of a sewer in Minneapolis, where some 3,000 American roaches were crowded in a space only twelve feet square. It is well known that roaches are sometimes able to enter buildings by way of sewers, but homes are probably much more often invaded via grocery bags.

Outside our circumscribed "affluent society" cockroaches live essentially everywhere with man. Roth and Willis mention that 2,500 roaches were captured in a single night in an African hut. The owner of a small fruit store in Puerto Rico is said to have cleaned out over a bushel of Madeira roaches. The problem of controlling roaches in places where food is handled is acute throughout the world, for here one cannot be lavish with highly toxic materials; and when food is abundant these insects have a way of reappearing quickly, as we have seen. The use of natural enemies to control roaches appears to have few possibilities, unless those relatively useless companions of man, dogs and cats, can be trained to feed on them. Animals that regularly feed on roaches include certain spiders, scorpions, wasps, toads, and hedgehogs, but a well-kept kitchen tolerates none of those interesting creatures.

This is why there is much interest at present in the possibilities of using the cockroach's own pheromones and hormones against him—or even better, against *her*. Failing here, we may have to resort once again to some of the interesting nonchemical control methods of an earlier time. Frank Cowan, in his fascinating book *Curious Facts in the History of Insects* (1865), mentions this ancient remedy: hold up a looking glass in front of the roach, and he "will be so frightened as to leave the premises." An even more remarkable practice, said to be in vogue at that time, is described as follows:

"It is no other than to address these pests a written letter containing the following words, or to this effect: 'O, Roaches, you have troubled me long enough, go now and trouble my neighbors.' This letter must be put where they most swarm, after sealing and going through with other customary forms of letter writing. It is well, too, to write legibly and punctuate according to rule."

Under the circumstances, it is remarkable that only a handful of people throughout the world are making a real effort to understand the cockroach. There are some 3,500 living species, but fewer than 5 per cent of these have been studied in any detail. What we do know suggests that every species is a story in itself and that even our best known species have still to yield final answers on many details of body function.

When a scientist is asked what good his research is, the classic answer (and a good one) is a shrug of the shoulders. A more thorough knowledge of roaches may or may not help us to reach a "peaceful coexistence" with them. But to a student of roaches, it is self-evident that any creature so beautifully adapted and adaptable for lo these millions of years is worth lifetimes of study. Not that we should all emulate roaches (though it might be exciting for a while if we could do so without selling our souls for a whiff of pheromone); but if there are any underlying principles of long-term survival, surely they are evidenced by the roaches. The study of roaches may lack the aesthetic values of bird-watching and the glamour of space flight, but nonetheless it would seem to be one of the more worthwhile of human activities. In fact, as I scan the evening paper, I wonder if it may not be more worthwhile than most of them.

Water Lizards and Aerial Dragons

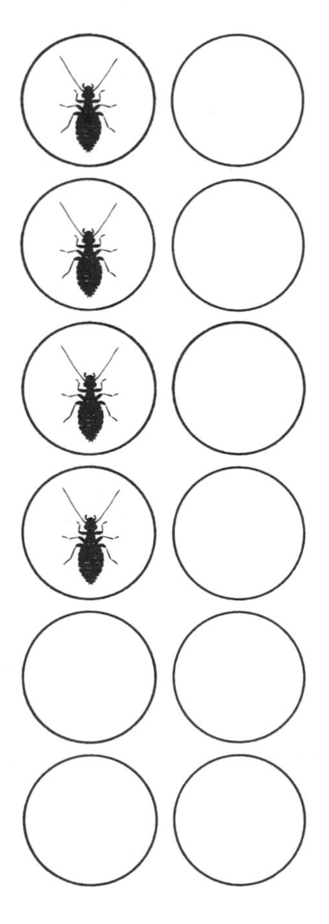

A man is as rich as the images collected in the recesses of his mind, images that flicker about in dreams or can be called upon to relieve an hour of stress or idleness. I shall let you in on one of mine: a scene along Pewter Pot brook in Hockanum, Connecticut, in the late 1920's. Sandbags are heaped across the stream to produce a pool of brown water in which occasional chunks of cattle dung can be seen floating, but the naked youths reveling in the water are not at all squeamish about such things. Crayfish dart about in the stream, redstarts in the willows, and the laughter of boys seems in keeping with the ebullience of nature. Blue-bodied, dark-winged damselflies patrol the stream, and high above, giant green darners make their entrances and exits. As a mind image, distilled by time, it is all a bit like an animated Japanese print, done in elegant colors and with a fine sense of balance. I hope that others who shared our swimming hole recall it now and then, too, for Pewter Pot is now a foul sluice through an indifferent chunk of Suburbia, and the image will never be recaptured there.

Dragonflies still delight me, not only because of the intimations of childhood they evoke, but because, as a professional entomologist, I recognize their uniqueness: in the remarkably varied world of living things, there is nothing even remotely like a dragonfly. As youths we called them sewing needles, though I don't recall that we were much concerned that they would sew up our ears as they were said to do. Others have called them devil's darning needles, snake doctors, horse stingers, mosquito hawks, and a variety of other things. Nowadays we seem to have settled on the word "dragonflies" for kinds that hold their wings out flat, damselflies for the more delicate forms that are able to elevate their wings vertically. In the early bestiaries they were called "libellae," a name that has come down to us as Libellula, the generic name for some of our commonest and most attractive dragonflies. It was once supposed that they took their name from their poise or balance on the wing, *libella* being the Latin word for a carpenter's level or a balance. But Lieutenant Colonel F. C. Fraser, retired from Her Majesty's service, and a world authority on dragonflies, traced the word back to 1554, when it was applied by Rondeletius to the hammerhead shark, and a year later to the larva of a damselfly, which is similarly "hammer-headed" and "fishtailed." Apparently the heads of these animals suggested the arms of a balance to Rondeletius. It was not until much later that these mud-dwelling larvae were known to be immature damselflies. In Thomas Mouffet's *Theater of Insects,* published in London in 1658, the larvae and adults were discussed in quite different sections, the larvae being called "lizards" which "lyeth in wait to catch fish." It must have been a shock when the world discovered that these ugly creatures transformed into something as spritely as dragonflies.

The sewing up of ears is not the only old wives' tale associated with dragonflies. The term "snake doctor" apparently arose from the belief that dragonflies warn snakes of approaching danger. According to Lucy W. Clausen, in her book *Insect Fact and Folklore,* there is a legend on the Isle of Wight that when boys go fishing, the dragonflies will guide good boys to the pools in which the fish are lying, but bad boys are stung. Many persons suppose that dragonflies can sting, but this is by no means true; they can be safely picked up with one's fingers so long as one does not come too close to their formidable jaws. In Japan, dragonflies have an important place in literature and art, and are considered symbolic of victory in battle. Japan is now one of the leading centers for the scientific study of these insects, and the Japanese word for dragonfly, *Tombo,* is the name of the only journal in the world devoted to this group of animals.

In parts of Asia and the East Indies, dragonflies (like certain other insects) have long been esteemed as food. The nineteenth-century naturalist A. R. Wallace reported that natives of the island of Lombok send their children out into the rice paddies with long poles smeared at the top with a sticky substance. When dragonflies alight on these poles, they cannot escape and are collected and served fried with onions. On the neighboring island of Bali, the natives are said to prefer them fried in coconut oil with vegetables and spices. In Thailand, Laos, and other parts of eastern Asia, dragonfly larvae are strained from streams and ponds, and served roasted. They are said to taste like crayfish.

We in the Western world have never allowed ourselves to acquire a taste for insects, although we consider as delicacies creatures considerably less "clean living" than most insects: lobsters and oysters, for example. Dragonflies represent a source of high-protein food we may yet learn to exploit. They are, of course, among the largest of insects. Our common green darners have a wingspan of nearly four inches, and the largest known living species, which occurs in South America, has a wingspan of seven and one half inches. As every well-informed schoolboy knows, there was a time, well before the reign of dinosaurs, when dragonflies with a wingspan of nearly thirty inches patrolled the air, the largest insects ever to have lived. These great creatures were basically very similar to modern dragonflies, and it is assumed that their larvae were also similar, though in fact no one has yet found a fossil larva, and we cannot be sure that they lived in the water like those of modern dragonflies. The full-grown larvae of these giants may have been nearly a foot long!

Considerably smaller dragonflies were common throughout the Age of Dinosaurs, and some remarkably well preserved fossils are known to us. Even today, there is little evidence that these insects are a declining race. Philip S. Corbet, of the Entomology Research Institute, Belleville, Canada, concludes his splendid book *A Biology of Dragonflies* with the remark that "dragonflies, their long history untarnished by defeat, still remain—monarchs of all they so completely survey." He attributes their continued success to their size and agility and to the fact that they have tapped a virtually inexhaustible source of food: the myriads of flies and other small insects that inhabit the earth. He also notes that their behavior is a good deal more flexible and adaptable than insects are usually given credit for. When conditions are unfavorable, for example, most dragonflies and their larvae are able to undertake one or more alternate courses of action. It will pay us to take a closer look at these versatile, successful, and highly decorative fellow inhabitants of our world.

Perhaps the most striking features of dragonflies and damselflies—aside from their two pairs of intricately netted wings, which may be clear and iridescent or variously banded with brown, silver-white, bluish, or various shades of red or yellow—are the great, protruding eyes that occupy the sides of their heads. As in all insects, these are compound eyes, that is, they are made up of a great many individual units, each with its own lens, or facet, surmounting the visual and nervous components. Some of the larger dragonflies have more than 28,000 facets in their eyes, some of these facing directly upward, others downward, forward, or even backward. This situation, along with the fact that the head can be rotated readily on the slender neck, makes it almost impossible to sneak up on a resting dragonfly without being seen. These wonderful eyes serve the dragonflies well in detecting and capturing small insects on the wing. It is said that some dragonflies are able to perceive movements more than forty yards away—most unusual for fixed-focus eyes such as those of insects. The antennae, which are conspicuous organs in most insects, are so small that they can barely be seen with the naked eye. This suggests that dragonflies are among the few insects in which the sense of sight is greatly dominant over smell and touch. This is borne out by observation and by experiments involving removal of the antennae, for antenna-less dragonflies navigate quite normally and even capture prey without difficulty.

Even more remarkable is the thorax, which is skewed in such a way that the legs are thrust way forward, while the wings are situated behind the hindmost legs, even though the wings properly belong with the same body segments as the second and third pairs of legs. If one examines the legs, he finds them all in a cluster just behind the mouthparts. They are covered with spines, and are so arranged that they form a basket for catching prey, which is then transferred to the mouth. Dragonflies are so highly adapted as aerial predators that they no longer use their legs for walking, although they are able to cling to plants and to some extent to climb about on them.

As flying machines, dragonflies differ in many ways from other insects. Because of the skewness of the thorax, the flight muscles are oblique, in fact sometimes more nearly horizontal than vertical. Furthermore, these muscles attach directly to the wing bases, one set on either side of a fulcrum, in such a way that the wings move over only a rather small arc. In most insects the major flight muscles attach not to the wing bases but to the strong, arching plate on the back between the wing bases; when these "indirect" flight muscles contract, they depress the top of the thorax and so press down on the wing bases and cause elevation of the wing. In dragonflies, the wings are attached rather close together on the back, and the indirect flight muscles play no important role in moving the wings. The direct flight muscles, which attach to the wing bases themselves, are, however, far better developed than in any other group of insects. The front and hind wings of dragonflies work independently

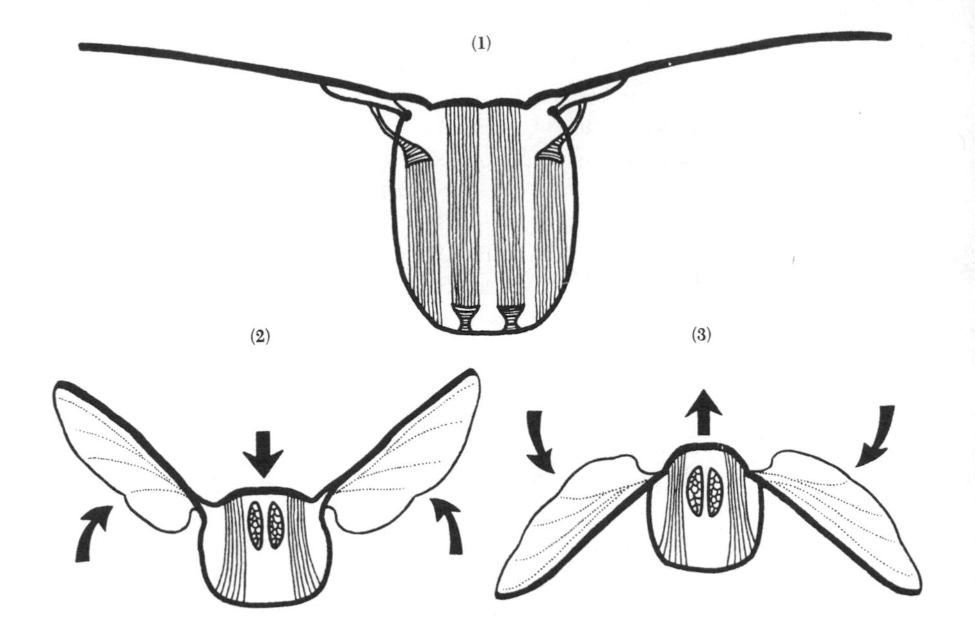

Flight mechanism of a dragonfly (*1*) compared with that of a fly (*2* and *3*). The major flight muscles of the dragonfly are attached one on each side of the wing base, and alternate contractions cause wing movements over a narrow arc. In the fly, major wing movements are produced by changes in the shape of the thorax brought about by muscles not attached to the wings at all. (Modified from R. E. Snodgrass.)

of one another—a point more easily confirmed with damselflies, with their characteristic slow, flickering flight. Furthermore, the wing muscles of these insects are of the synchronous type, which means simply that the wings beat once in response to each nerve impulse, at a rate of up to about 30 beats per second. This stands in strong contrast to more recently evolved insects such as flies and bees. The honeybee may beat its wings at a rate of 200 to 250 times per second, while certain midges are reported to have a wingbeat frequency of over 1,000 per second. This much faster rate is possible because these insects have asynchronous flight muscles, that is, the muscles have their own very rapid rhythm of contraction which does not exhibit exact correspondence with the nervous stimuli received. It is said that weight for weight the flight muscles of higher insects generate more energy than any other living tissue.

But dragonflies are limited to a relatively slow wingbeat frequency, to a small arc of movement and lack of synchrony between the front and hind wings, and to the use of strongly slanted, presumably "primitive," direct flight muscles. Furthermore, they have no mechanism for flexing the wings backward so that they can lie flat on the abdomen, an art mastered long ago by even such ancient creatures as the roaches. Because of their many archaic flight features, entomologists speak of these insects as "palaeopterous" (ancient-winged), and regard them as relics of a very early stage in insect evolution. Presumably the insects went through a stage when they simply glided from trees, a bit like flying squirrels. Since gliding had obvious advantages in escaping from predators and in dispersal, natural selection favored expansions of the plates on the back so as to support them for longer periods on air currents. The tendency for insects to flex their legs upon landing may have favored development of stronger leg muscles, and since these muscles normally attached to the sides of the back, they may have eventually proved capable of moving the gliding surfaces also attached there. With the development of a definite fulcrum of movement, a thinning of the flaps to transparent membranes supported by a network of braces, and refinement and elaboration of muscles already present in the thorax as direct flight muscles, a flying machine a bit like a dragonfly may have arisen. Not every stage is documented by fossils, but dragonflies (and to a lesser extent mayflies and some other primitive insects) are surely descendants of some of these early experiments in flight.

All this seems to suggest that dragonflies don't really fly very well. Anyone who has never attempted to catch a dragonfly on the wing should give it a try; it is quite a sport, with the dice heavily loaded for the dragonfly. Dragonflies may not be the fastest of insects, but they do pretty well (up to about thirty miles per hour). However, their ability to maneuver, enhanced by their all-seeing eyes, is not excelled by any other flying creature. Damselflies are admittedly weaker fliers, but even they are a good deal more difficult to capture than they appear. R. J. Tillyard, a brilliant Australian entomologist who was a foremost authority on dragonflies before his death in an automobile accident in 1937, summed it up well when he remarked that these insects form "a singularly isolated group, marked by very high specializations of structure, superimposed upon an exceedingly archaic foundation."

Dragonflies are also capable of traveling great distances at times. A large flight of a species called the "globetrotter" was once observed by persons on a ship nine hundred miles off the coast of Australia, from where the dragonflies were presumably coming. This species, like several others, breeds in temporary pools, often in regions of irregular rainfall, and as an adaptation to this precarious environment the larvae develop rapidly, and the adults are able to move great distances and find places where there are fresh pools. It is characteristic of dwellers in temporary ponds that they have unusually broad hind wings. When they are ready to migrate, they apparently launch themselves high into the air, where they are borne great distances by air currents on their broad wings, apparently keeping together in swarms by vision. Since convergent wind currents generally lead to heavy rainfall, the dragonflies are carried into regions soon to have an abundance of pools suitable for breeding.

W. H. Hudson, in his book *The Naturalist in La Plata,* described large flights of dragonflies that appear in advance of seasonal winds, or *pamperos,* on the plains of Argentina. "They make their appearance from five to fifteen minutes before the wind strikes; and when they are in great numbers the air to a height of ten or twelve feet . . . is all at once seen to be full of them, rushing past with extraordinary velocity in a northeasterly direction. . . . Of the countless millions flying like thistledown before the great pampero not one solitary traveller ever returns." Hudson observed the dragonflies spending the night in protected places, "clustering to the foliage in such numbers that many trees are covered with them, a large tree often appearing as if hung with curtains of some brown glistening material, too thick to show the leaves beneath."

Our common green darner, Anax junius, migrates north into Canada each summer, and its progeny fly back south in the fall, much in the manner of the monarch butterfly. Mass southward flights in September and October typically occur following passage of cold fronts. Apparently the dragonflies, like monarch butterflies and many birds, take advantage of tailwinds and thermal updrafts. At Hawk Mountain Sanctuary, in Pennsylvania, thousands of green darners are sometimes seen moving south in the fall—and are sometimes seen being preyed upon by hawks, which are among the few natural enemies of the larger dragonflies.

Even nonmigratory species often fly considerable distances in their search for prey. Large numbers of dragonflies, sometimes of several species, occasionally congregate in places where there are swarms of midges or other small insects. At these times the dragonflies fly close together and show none of the aggression that they often display at their breeding sites. Some dragonflies will follow large grazing animals about and snatch up the insects the animals stir from the grass or the flies that swarm about them. No one who has watched dragonflies darting about in a swarm of insects can doubt that, "palaeopterous" though they may be, they make incredibly good use of the equipment they have. It is worth remembering that the insects they prey upon often are the very same ones hailed by entomologists as the best fliers. It would little console a midge passing down the gullet of a dragonfly to know that its highly developed indirect flight muscles are capable of a thousand contractions per second, compared to which the dragonfly is a piker. The honeybee, that ultimate of insects, is sometimes preyed upon by large dragonflies that dash back and forth in front of the hive entrance and snap up the bees one by one as they come out. All of which means that words such as "archaic" and "primitive" can't be equated with "unsuccessful." Many a wonderful beast has gone down the road to extinction while dragonflies and roaches thrive as of yore, and the lowly horseshoe crab and other ancient sea animals remind us of even earlier times, when there was no life at all on land.

Despite the fact that dragonflies often fly considerable distances in their hunting flights, and some of them hundreds of miles during their migrations, the focal points of their lives are the ponds and streams in which they breed. Their attraction to bodies of water is, again, visual. This is shown by the fact that they occasionally patrol shiny surfaces such as paved roads or show sexual behavior over the roofs of automobiles, sometimes even going so far as trying to lay eggs in these ridiculous situations. This seems stupid indeed until we recall that roads and automobiles are hardly part of the world in which dragonflies evolved. As a matter of fact, many dragonflies appear to check the evidence of their eyes by "water touching"; that is, after arriving at a reflecting surface they dip their abdomen in briefly, a trait that doubtless serves them well in a world often strangely altered by man.

At the La Brea tar pits in Los Angeles, dragonflies have sometimes been seen dipping their abdomens into shiny pools of viscid petroleum, only to have them stick there. Sometimes, with a violent beating of wings, the dragonfly is able to free itself, only to try once again. The bodies of dead dragonflies are often seen floating on the pools. Even though to human nostrils the tar pits reek with the odor of crude oil, the dragonflies either do not detect the odor or do not respond to it.

All dragonflies and damselflies are more or less selective as to the type of situation in which they breed. Many damselflies inhabit small streams, laying their eggs inside plants where they will be protected from the turbulence of the stream. Some tropical damselflies have very long abdomens, which they use for thrusting their eggs into the water held among the leaves of epiphytes growing on trees in the rain forest. Dragonflies, for the most part, inhabit more open water. Some species choose large rivers or the shores of lakes; others prefer sluggish streams, weedy ponds, or even swamps; still others utilize temporary pools or even, in a few cases, lay their eggs in depressions that are likely to be flooded some time in the future. Many dragonflies lay their eggs loosely in the water by simply dipping the abdomen beneath the surface and washing them free. This behavior is easily observed on a summer day at almost any body of water that has a population of dragonflies.

In fact it is generally the males that arrive at the breeding sites first.

In many cases, the males become localized over a certain part of the pond or stream, and return to the same place day after day, defending this territory against intrusion by other males. Damselflies and small dragonflies often establish very limited territories close to vegetation or to the water surface, while larger dragonflies are "stacked" in larger territories above them. Norman W. Moore, of the British Nature Conservancy, made an interesting study of the dragonfly population of a flooded bomb hole in Dorset. Close to the water there were twenty to thirty-six small damselflies, somewhat higher up six to eleven small dragonflies, still higher two rather large Libellulas, and finally a single giant darner patrolled the entire hole from a considerable height. In some cases the number of individuals in an area also influences territory size; thus a single male may patrol a whole small pond, but if two or three are present they tend to divide the pond into halves or thirds, as the case may be. Since the females tend to space themselves in accordance with territories laid down by the males, it has often been assumed that territoriality serves the same function as it does in birds: the prevention of overconcentration and hence overexploitation of the habitat.

In some species, the extent to which a male has to defend his territory tends to be correlated with population density—the more males, the more contacts between them. As with most territorial animals, physical combat occurs only as a last resort. This is especially fortunate in the case of dragonflies, which are quite capable of tearing each other apart but unable to heal any of their wounds. There might well be a fantastic sacrifice of males if dragonflies had not evolved certain symbolic aggressive displays. V. I. Pajunen, of the University of Helsinki, discovered that males of one of the common small Finnish dragonflies survey an area seven to thirteen feet wide surrounding their perches, and permit no other males to enter. Males of other species, or nonaggressive males of their own species, are usually merely pursued a short distance. However, when an aggressive male of the same species enters, a complex threat display follows. The two males dash toward each other, pass without making contact, then turn about and dash toward each other again. They do this repeatedly, all the while losing altitude, so that they describe a spiral. This "spiral threat display" usually results in the departure of the intruding male, but if it does not, actual fighting may ensue. According to Professor Pajunen, fighting "is obviously a modified form of preying behavior, including grasping with the legs and probably also biting." Following either fighting or threat display, the victorious male chases off the loser for a distance of several feet. Some of the larger dragonflies will also attack males of other species, producing a noisy clash that results in one of the males flying off with the other in pursuit. In some cases large male dragonflies actually devour smaller species.

Some of the bright colors of male dragonflies are associated with threat displays. The males of one of the commonest inhabitants of ponds in North America, Plathemis lydia, have the abdomen bright silvery white above. Merle Jacobs, of Goshen College, Indiana, has found that these conspicuous white banners play an important role in male interactions:

"When two males encounter one another, one male dashes at the other and pursues him. The pursuing male raises the white upper surface of the abdomen toward the new arrival, while the latter flies away with abdomen lowered. . . . After the second male has been 'chased' 8–16 m. [9–18 yards] from the site over which the first had been active, the first usually turns and flies back with abdomen lowered while the second now pursues with abdomen raised. This pursuit display, alternating in direction, continues until the new arrival restricts his movements to the vicinity of another site."

Dr. Jacobs captured sixty-eight males and painted their abdomens black. These sixty-eight males had a significantly lower average "mating score" than did an equal number of unpainted males, chiefly because they were less successful in defending territories. He also applied white paint to the normally dark abdomen of several females. Surprisingly, they were still accepted by the males as females, probably because of their manner of flight and different wing pattern. However, the females did sometimes treat these white-tailed females as males, suggesting that different elements in the color pattern are utilized in sex recognition.

Experiments on the elements that elicit sexual behavior were performed by Christiane Buchholtz of the Zoological Institute of Munich, Germany. She presented various male damselflies with models suspended from the ends of fish poles. In one case, she found that the model must consist of at least the form of the head and thorax and one wing; the thorax must have the green-and-black pattern characteristic of that species, and the wing must be clear and translucent. Curiously, wing shape makes no difference, and the addition of other wings of various colors also makes no difference so long as one wing is suitably transparent. However, the models have to be moved to and fro with a certain frequency in order to induce courtship by males. In one complex of several closely related species of damselflies, she found that sexual isolation among the species is maintained largely by the transparency of the wings of the female. In one form, the males court females with wings allowing at least 30 per cent of the light to pass through, while in another form the percentage is 20 to 22 per cent, in another only 8 per cent.

The damselflies of North America are now being subjected to a detailed, comparative study by George H. Bick, of Saint Mary's College,

Notre Dame, Indiana, assisted by his wife, Juanda. The Bicks have found that several species show little or no territorial behavior and virtually nothing that could be called "courtship": the male simply seizes an approaching female, with no apparent exchange of behavioral signals. Yet the males are well able to recognize the females of their own species, apparently by quick recognition of elements in their color pattern. Males of a few species are quite definitely territorial. The Bicks found that in one of our common damselflies, Argia apicalis, the males space themselves at about six-foot intervals, defending their territories by flying at intruders and by a characteristic wing-flicking called "wing warning." By these ritualized movements they are able to win nearly 90 per cent of their encounters—with their own or other species—without resorting to actual combat. The effect on some of the other, less aggressive species is to squeeze them into niches not occupied by Argia apicalis.

These observations suggest that territoriality may be important in enabling various species to "divide up" the available resources: that is, that it serves an important role *between* species, and not necessarily only *within* species. Several workers have noted that when the population of one species is very high, territoriality may break down, and the males may show a much higher degree of social tolerance than they ordinarily would. If this is the case, can territoriality be said to help in preventing overpopulation, as I suggested earlier? What it may do, when populations are low to moderate, is to ensure that, by wide spacing, some individuals are forced to explore other locations, thus leaving no suitable places unoccupied. In other words, territoriality may serve less to prevent overcrowding than to prevent under-utilization of the available resources.

Many dragonflies and damselflies undergo striking changes in color during their adult lives of only a few weeks. For the first week or two after emergence they may appear relatively drab. Then, as they enter their reproductive period, the males acquire areas of whitish or bluish pruinescence, suggesting the bloom on a plum. Later in life this tends to wear off. A male during reproductive activity may look very different from a very young or very old male, so much so that he might readily be thought to represent a different species. Adding to the complexity is the fact that the males and females of one species often appear very different, the females not only lacking the pruinose areas but having different wing patterns and colors. The "ruby spot" damselflies of tropical and temperate America are among the most vividly colored of all, the males having the wing bases bright red, although in the females these areas are merely dull brown. In the East Indies there are small dragonflies in which the wing bases of the males are brilliant, metallic blue-green.

Examples such as these did not escape the attention of Charles Darwin when he developed his theory of sexual selection: that "the more attractive individuals are preferred by the opposite sex." The importance of color in the lives of damselflies accounts for the fact that they are sometimes attracted to bright spots of paint or to brightly colored floats on fishing lines. All this is in striking contrast to the odor world of the cockroaches we met in the previous chapter. The world of dragonflies and damselflies is a world of colors, shapes, and movements. That we, too, are "visual animals" doubtless accounts for our estimation of these insects as things of beauty, while roaches are mentioned with disgust. Perhaps if we had small, inefficient eyes and great, powerful noses we would find roaches more attractive. Or would we?

In dragonflies colored patches provide important signals for sex recognition or threat displays. In damselflies they often play specific roles in courtship. In some species, after a male recognizes a female in breeding condition, he settles on a stem, spreads his wings slightly, and curls his abdomen so as to expose a bright spot on its tip. In other species there are prominent white spots on the legs; in this case the male flutters back and forth in front of the female, facing her and dangling his brightly colored legs. Many, many times we simply do not know the precise function of the varied colored spots and bands of damselflies and dragonflies. A thing of beauty is a joy forever; but a thing of beauty explained rewards both eye and mind.

The actual mating act of Odonata is one of the most bizarre performances to be seen anywhere. Doubtless if a trip to the Kalahari Desert were required to observe it, more publicity would result. But the fact is that a few hours—or even a few minutes—spent in quiet observation by pond or stream in the summer is all that is needed. Before discussing mating, it is necessary to describe the abdomen briefly (for in our enthusiasm for the eyes and wings of dragonflies, we have never reached the abdomen, even though some of the abdominal structures are strangest of all). There is nothing special about the female, the long, needle-like abdomen merely terminating in the sexual orifice and the egg-laying apparatus. In the male, however, the abdomen terminates in a double set of powerful clasping organs—not unusual in itself until one realizes that these organs are not designed for grasping the genital organs of the female but for grasping her neck. In fact, many female dragonflies have grooves and pits on the back of the head into which these claspers fit. Here then is a dilemma: the claspers (and genital opening) of the male are located at his tail end, but this is applied to the neck of the female rather than to her genital opening. The dilemma is solved by the possession, in these and in no other insects, of a special genital pouch beneath the forward segments of the male abdomen—actually well in front of the middle of the body. Here one finds a jointed

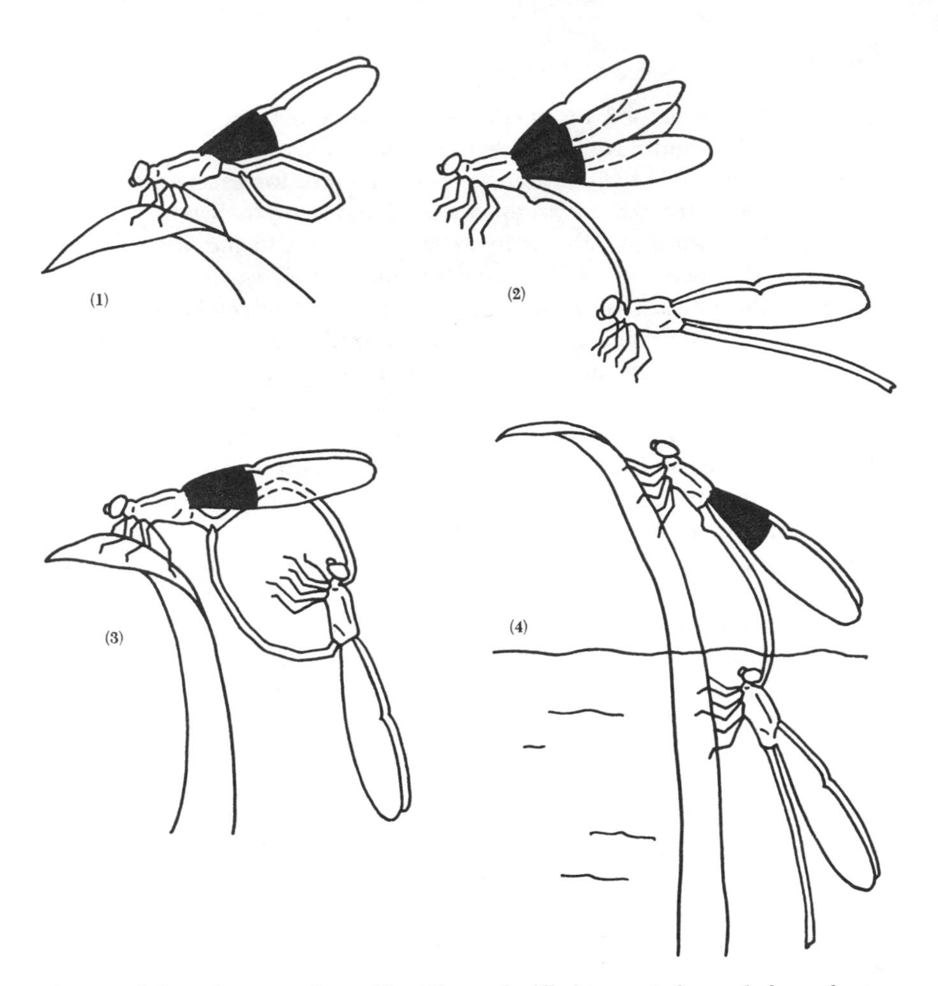

Mating of the ruby-spot damselfly. The male fills his genital pouch from the true sex organs at the end of his abdomen (*1*). He then courts a female and eventually grasps her by the neck (*2*). She loops her abdomen about and applies the tip to his genital pouch (*3*). Following sperm transfer, the male continues to accompany the female, even as she backs into the water to lay her eggs (*4*). (After Clifford Johnson.)

penis and a variety of hooklike appendages associated with a sac that has no connection with the testes at all. Thus the male must fill these secondary genital organs, which he does by making a loop of his abdomen, applying the tip briefly to the underside of the base, and permitting sperm to enter the sac at the base of the penis. In damselflies, the sac is often filled after the male has grasped the female, but most dragonfly males are "loaded" before they pair with a female.

The act of copulation is initiated by the male following slightly above the female, then seizing her thorax with his legs. At this point, if the female is receptive, she flies slowly and allows herself to be carried. The male then makes a loop of his abdomen and grasps the female in the neck region with his claspers. He then lets go with his legs, straightens his abdomen, and flies ahead, carrying the female in tandem. The female then normally makes a loop downward and forward with her abdomen and applies her genital opening to the genital sac of the male. This is called the "wheel position," since in fact the two form a loop, the female mostly upside-down beneath the male. The pair may remain together for anywhere from three seconds up to an hour or more, but all species that require more than a few seconds copulate while perched.

Several persons have attempted to postulate some of the intermediate stages by which this unique system of mating may have evolved, but it is all rather like a Chinese puzzle. In all probability the dragonflies evolved from a very primitive type of insect (perhaps like the bristletails mentioned in Chapter 2) in which the male deposited a sperm droplet on the ground to be picked up by the female. But with the acquisition of flight, it became more practicable for the male to deposit the sperm on his own body, and the base of his abdomen would be the place most easily reached. Perhaps at first the sperm droplet was merely fastened to

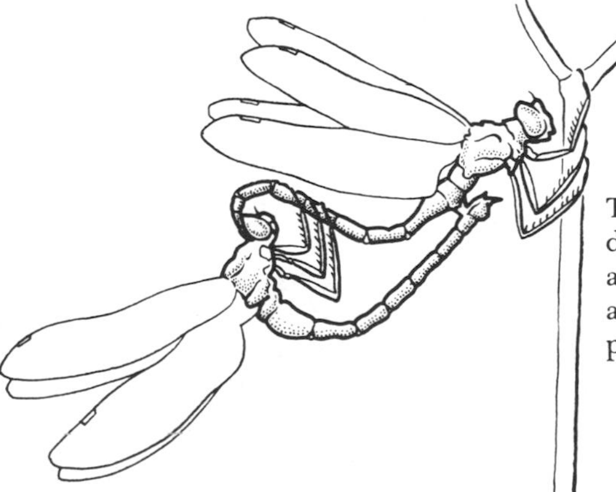

The wheel position of mating dragonflies. The female is below and has applied the tip of her abdomen to the male's genital pouch. (After P. P. Calvert.)

the surface of the body at this point, and later the complex secondary genital structures evolved. This is the hypothesis preferred by Corbet in his book *A Biology of Dragonflies,* and it is perhaps the best that can be done. We know that dragonflies have copulated in much this manner since the Age of Dinosaurs, but the fossils do not tell us very much about the intermediate stages.

This is still not the end of this remarkable story, for in most cases the male continues to grasp the female after mating and actually leads her to a spot suitable for laying her eggs, often a different place than he had defended previously. In the case of the ruby-spot damselfly, mentioned earlier, the male leads the female, still holding her by the neck with his abdominal claspers, to the stem or leaf of an aquatic plant. There he releases her and perches facing her while she goes beneath the water to a depth of three to five inches and lays her eggs in the stems of plants. In other damselflies the male continues to hold the female in tandem during the process of egg-laying, the two sometimes descending well beneath the surface. In the case of our common white-tailed dragonfly, Plathemis lydia, the male releases the female after mating, but guides her to a clump of sticks or mud, then hovers above her while she lays her eggs just beneath the surface. If other males approach the female at this time, they are driven off summarily. If another female approaches, however, the male may copulate with her and place her near his first spouse and guard them both. Despite the intense rivalry among the males, the females appear content to lay their eggs side by side.

Dragonflies often mate several times, and the females of most species lay several batches of eggs, but the water world in which the larvae develop is filled with predators, so only a small percentage of the eggs normally produce adults. The length of the larval stage is extremely variable. Dragonflies that breed in temporary ponds often pass through their larval stage in only a few weeks, while some inhabitants of permanent waters require as much as five years. The majority of species take the better part of one or two years.

Compared to the adults, the larvae of Odonata are ugly indeed, but in their own way they are just as remarkable. Some species live in the mud of the bottom and tend to be squat and hairy, while others climb about on aquatic vegetation and are much more streamlined. The larvae of damselflies are readily distinguishable from those of dragonflies, since they have three long, usually rather flat gills extending from the tail end of the body. The larvae of both damselflies and dragonflies tend to be colored so as to match their background, and they are able to change color to some extent when the background changes. Some larvae that crawl on plants are green in the summer and brown in winter. Others are banded when young, presumably to break the profile and render them less conspicuous among the weeds, assuming solid colors when they are larger and less subject to predation (in fact, larger larvae of the same species may be their worst enemies). Young, banded larvae of certain dragonflies, if removed from the checkered pattern of light in the weeds of their habitat and placed on a solid-black background, tend to become all dark much sooner than they would if left to mature in nature.

The larvae of dragonflies, though not those of damselflies, are well known for their ability to move by "jet propulsion," a phenomenon easily studied in an ordinary kitchen pan. These larvae have an enlarged, muscular rectum, and breathe by forcing water in and out and extracting air through "rectal gills." When disturbed, they are able to contract the rectum suddenly, forcing out a strong jet of water that propels them forward several inches. A person who lifts one from the water suddenly must beware lest he be squirted in the eye. The larva of the green darner is said to be able to move at a speed of twenty inches per second. We know that dragonfly larvae have "giant fibers" in their nervous system much like those of roaches. Stimulation of some of the small hairs at the tail end of the larva causes a trigger-quick response: the legs are drawn in to the sides of the body; the rectum contracts violently; and the larva is propelled away from the source of stimulation, perhaps a predatory fish, water-beetle larva, or another dragonfly larva.

Dragonfly larvae themselves are among the most formidable predators in fresh water, and their prey-catching apparatus is, like so many features of these insects, unique. The lower lip is enormously lengthened, and has a double hinge joint so that it can be pulled back beneath the body when not in use; the outer part is expanded and provided with stout hooks, and in resting position forms a "mask" that covers much of the face of the larva. The lip is capable of being thrust forward suddenly, and the terminal hooks are capable of grasping prey well in front of the larva and pulling it back to the sharp, jagged mandibles. Dragonfly larvae prey upon many kinds of small insects occuring in the water, and the larger ones are well able to handle small fish.

The larvae of large dragonflies, such as the green darner, lie in wait for prey to pass close by. When an object is spotted, the larva turns to face it. The eyes are large and so situated that the larva has excellent binocular vision, and is well able to estimate the size and distance of the prey. As the larva creeps up on the prey, the image moves over facets closer and closer to the midline of the body. The optical axes of certain of the inner facets of each eye intersect at the point where the hooks of the extended lower lip will reach, and as soon as the image of the prey reaches this point the lip is shot out and the prey grasped. The larvae of some dragonflies, and especially those of damselflies, have small eyes and fairly long antennae, and these larvae depend less on sight and more on touch and on the detection of vibrations in the water.

The movement of the lower lip is exceedingly fast. According to Gordon Pritchard, of the University of Calgary, Alberta, the forward thrust requires less than three one-hundredths of a second, with an additional six to twenty hundredths of a second being required for retraction. This wonderfully quick reaction is believed to be produced by blood pressure. Apparently the muscles of the body wall contract suddenly, causing blood to rush to the head and force out the lip, although we still do not understand the mechanism fully. Dr. Pritchard's high-speed motion pictures of the process reveal that the anus of the larva is closed tight during prey seizure; otherwise some of the pressure might be directed backward and bring the "jet propulsion" apparatus into unwanted functioning.

Dr. Pritchard found that dragonfly larvae will often strike at objects that are separated from them by a sheet of glass; but they do not react, even when hungry, to extracts of crushed insects when they are released in water. Apparently they are as indifferent to odors as the adults. In many cases both touch and sight play a role in prey capture. The larva of Libellula, for example, if kept in total darkness with a variety of other aquatic insects, will eventually capture most of them. To test the relative importance of sight and touch in such larvae, Dr. Pritchard rigged up a small dummy with a slender thread attached. When the thread did not touch the larva, the latter always struck at the sight of the dummy, but when the threat was in contact with a leg, the larva struck at the point of contact in preference to the sight of the dummy. On the other hand, vision is all-important in active, big-eyed larvae such as those of the green darner. When such larvae are confined in total darkness with a variety of aquatic insects, they are able to catch nothing, even over a period of days.

Some years ago Cyril Abbott, of Pembroke State College, North Carolina, performed a number of experiments with green darner larvae in ordinary pans of water—experiments that ought to be easily repeatable, since these are among the easiest larvae to find. He found that larvae with both eyes blacked out are wholly unable to feed themselves, and soon starve to death. Larvae with only one eye blacked out often starve to death, too, since they ordinarily snap to the right or left of the prey. However, a few of those that Abbott studied adapted to monocular vision and succeeded in catching enough prey to survive. They did this by spotting the prey in the one good eye, approaching eye-forward by creeping crabwise, and then, at a certain distance, shooting out the lip and rotating at the same time, so that they suddenly faced the prey. This seems to indicate that a single eye, with its hundreds of facets, is capable

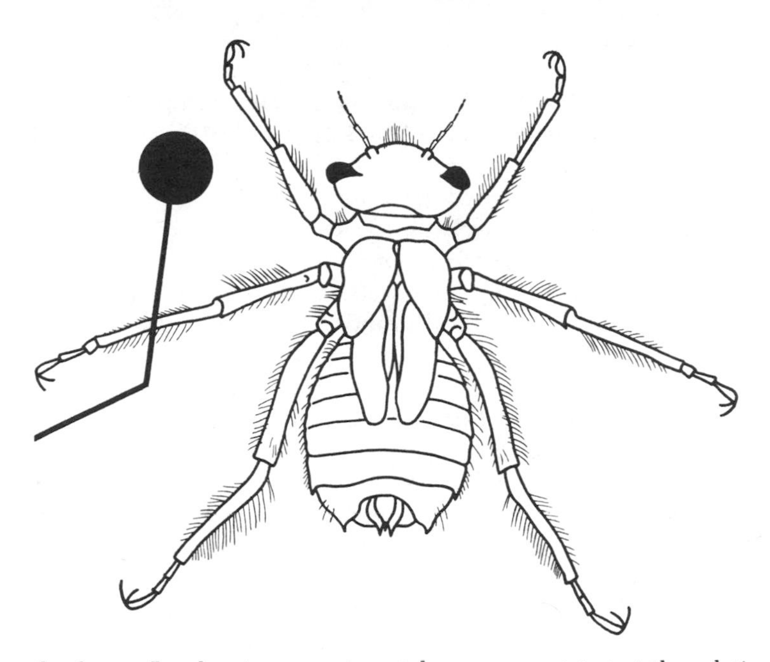

Larva of a dragonfly, showing experimental arrangement to test the relative importance of sight and touch. When the thread leading from the circular dummy touched the larva, it struck at the point of contact; when the thread did not touch the larva, the latter responded to the sight of the dummy. (After Gordon Pritchard.)

of at least crude binocular vision and that some larvae are capable of learning quite rapidly: still another example of the wonderful adaptability of dragonflies.

According to Professor Abbott, dragonfly larvae kept in pans are at first inclined to dash about madly or hide under an object when someone approaches. However, they soon become conditioned to the presence of humans, and after a few days of being fed from the end of a needle they actually turn toward the experimenter as if anticipating food. Abbott was able to entice some larvae out of the water for food. Another interesting bit of behavior can be observed in larvae taken from the water: they often lie motionless, with their legs drawn up close to the body, apparently feigning death or "playing possum." Abbott found that larvae could be induced to enter this trancelike state by gently stroking the sides of their abdomen—a reaction, as Philip Corbet notes, exactly the same as occurs in the African spitting cobra.

One of the problems of rearing dragonfly larvae in captivity is that of keeping them supplied with live insects as food. Philip P. Calvert, of the University of Pennsylvania, once titled an article "How Many Mosquito Larvae and Pupae Are Required to Make One Dragonfly?" A dragonfly larva that he reared consumed 3,037 mosquito larvae in the course of its life of about one year, as well as 164 mosquito pupae and a few other things, including 17 larvae of dragonflies and damselflies! Adult dragonflies also consume quantities of insects, of course, including many adult mosquitoes, although only locally and occasionally do they cause any noticeable decrease in the numbers of flying insects. Philip Calvert, by the way, was a person who let a youthful enthusiasm for dragonflies carry him to fame as a world authority on that group. His first article, published when he was twenty-two, concerned the dragonflies of his native city, Philadelphia, and when he died in 1961, at the age of ninety, he had published over 750 articles and books on his favorite insects.

The story of the transformation of the dragonfly larva to an adult has been told many times, though in fact few persons have observed it, since the larvae usually crawl out onto a plant or other object at night and molt in the very early morning. Last summer my family and I were lucky enough to watch a transformation right in front of us as we ate lunch on the shore of Fish Lake in Glacier National Park. It was a place we shall not soon forget, and that obliging dragonfly added a touch that etched it even more deeply in our memories. But for a description of the molt to the adult stage, I shall defer to a person who for many years shared with Calvert the distinction of being America's most articulate spokesman for dragonflies, James G. Needham of Cornell University. (A few years before his death, Professor Needham tried to convert me to dragonflies, but I had already cast my lot with the wasps.) In Needham and Westfall's *A Manual of the Dragonflies of North America,* the final molt is described in these words:

"First comes a splitting of the skin down the back and across the head, and the pushing up through the split of the head and thorax. The legs are slowly withdrawn from their sheaths. Soon a mighty effort lifts the front of the body and leaves it standing on its tail, so to speak, in the loosened skin of the abdomen. After a rest of a few minutes, with legs folded, the dragonfly makes a final lurch forward and seizes a footing; then, with withdrawal of the abdomen, the entire body at length stands free. It is a sorry-looking thing, misshapen, with tightly crumpled wings. . . ."

But by the time the sun is well in the sky, the wings are fully expanded and dry, and the body has hardened and assumed much of its color pattern. Soon the dragonfly departs on its maiden flight, propelled through a different medium by utterly different locomotory mechanisms and putting to use a prey-catching device entirely different from the one it had only a few hours earlier. The water lizard has become an aerial dragon, a creature that, to quote Thomas Mouffet (1658) "doth set forth Nature's elegancy beyond the expression of Art."

Corbet, Longfield, and Moore conclude their book *Dragonflies* with the comment that these animals have, in the final analysis, almost no effect on the economy of man. But they add: "Animals (especially dragonflies!) are valuable because they are beautiful." To this I can add only: Amen. So long as man is a thinking and feeling being, he will have a need for dragonflies; the student will ever find in them a source of fascinating problems, and the artist and poet in all of us will feel some of the joyous surprise of Tennyson when he penned his well-known lines:

> Today I saw the dragon-fly
> Come from the wells where he did lie.
> An inner impulse rent the veil
> Of his old husk: from head to tail
> Came out clear plates of sapphire mail.
> He dried his wings: like gauze they grew;
> Thro' crofts and pastures wet with dew
> A living flash of light he flew.

The Cricket as Poet and Pugilist

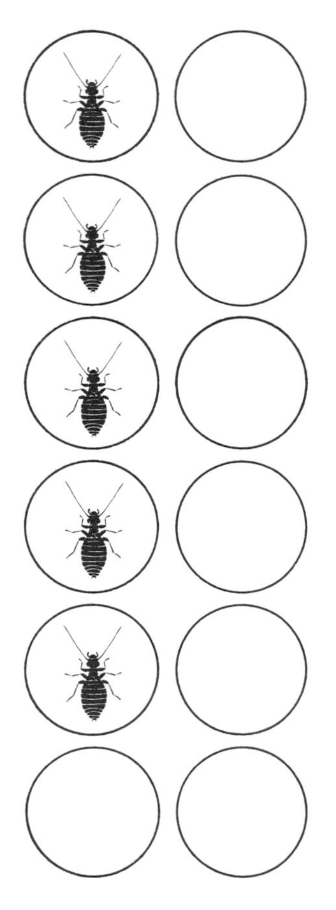

It appears difficult for man to develop a rapport with insects. Is it because insects are so different from us or because we have become irrevocably biased against creatures so small, yet capable of pricking our ego with their bites and stings and of seriously competing with us for the bounty of the earth? Biased we are, for most of us are brainwashed in childhood against anything devoid of hair or feathers: What good is a toad, a snake, a worm—most of all, an insect? Even a mature and thoughtful man, imbued with a reverence for living things, has not quite the same feeling for a beetle that he has for a squirrel. True, Robert Burns did write a poem "To a Louse," but it doesn't have quite the lofty sentiments of Shelley's "To a Skylark." Perhaps Maurice Maeterlinck summed it up best in these familiar sentences:

"The insect does not belong to our world. The other animals, even the plants, despite their mute existence and the great secrets which they nourish, do not seem wholly strangers to us. In spite of all, we share with them a certain feeling of terrestrial fraternity. They surprise us, even make us marvel, but they fail to overthrow our basic concepts. The insect, on the other hand, brings with him something that does not seem to belong to the customs, the morale, the psychology of our globe. One would say that it comes from another planet, more monstrous, more dynamic, more insensate, more atrocious, more infernal than ours."

Nevertheless there are a few insects that seem somewhat less alien to us: butterflies and lady beetles, perhaps, and certainly crickets. Jiminy Cricket has even become a motion-picture star. The cricket is surely a gentleman, neatly cloaked in brown or black, and minding pretty much his own business. And he sings, not quite like Caruso perhaps, but at least a pleasant, rhythmic lay that makes up in persistence what it lacks in variety. When a cricket is chirping, life is just a bit richer for all within hearing range. To me there is nothing quite so pleasurable as a September evening filled with the songs of crickets and grasshoppers. My pleasure has a touch of melancholy, true, for it will all be over soon; but at the same time I am awed by the fact that a perfectly ordinary chunk of countryside can house such an array of elegant vocalists.

The brown house cricket, originally a European insect, has become the basis of a small industry. Several thriving cricket farms scattered about our southern states supply bait for fishermen and experimental animals for schools and colleges. As pets, crickets are inexpensive, clean, and musical, even if they don't catch mice or frighten off burglars. Nowadays we hear little of the cricket on the hearth, so dear to John Milton and to Charles Dickens; in fact the hearth is no longer the utility or the symbol it used to be. But Cowper's verses still remind us of perhaps the closest rapport man and insect have ever enjoyed:

> Little inmate, full of mirth,
> Chirping on my kitchen hearth,
> Whereso'er be thine abode
> Always harbinger of good,
> Pay me for thy warm retreat
> With a song more soft and sweet;
> In return thou shalt receive
> Such a strain as I can give.
>
> Thus thy praise shall be expressed
> Inoffensive, welcome guest! . . .
> Frisking thus before the fire,
> Thou hast all thine heart's desire . . .
> Wretched man, whose years are spent
> In repining discontent,
> Lives not, aged though he be,
> Half a span, compared to thee.

John Keats dedicated one of his finest sonnets to the cricket and the grasshopper, beginning with the immortal line "The poetry of earth is never dead." Alas, the Romantic Age is long since gone. We talk of the productivity of the earth, not its poetry. And the cricket has been psychoanalyzed. He has turned out to be a less simple soul than the romantics believed (but then, so has man). At the same time he has turned out to be a good deal more exciting.

The singing of crickets has intrigued man for a very long time. The very word "cricket" is onomatopoetic, being based on the same stem as the word *creak;* the French equivalent is *crequet,* the Dutch *krekel.* When the American Association for the Advancement of Science met for the first time in Philadelphia in 1848, the great Louis Agassiz rose to address the audience not on his favorite subjects of fossil fish, embryology, or glaciation—but "On the Phonetic Apparatus of the Cricket." "Prof. A.," the account reads, "proceeded to demonstrate the structure of [the] wings on the black-board." Doubtless the audience was captivated by the Agassiz magic, but there is no evidence that he presented anything not already pretty well known at the time. The British entomologists William Kirby and William Spence presented a brief account of sound production in crickets in their book *An Introduction to Entomology,* which went through several editions between 1815 and 1828.

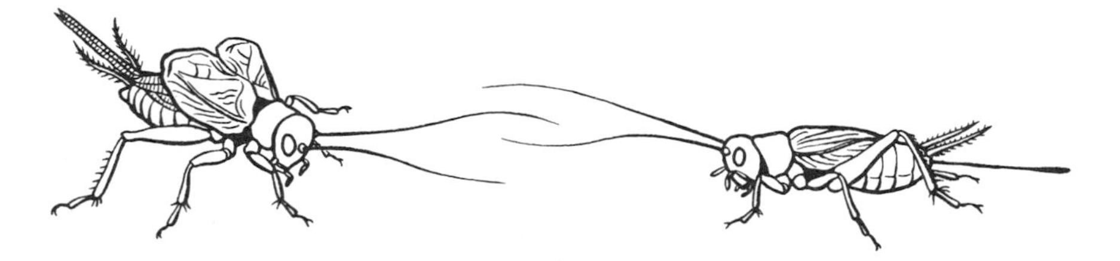

The courtship of the field cricket. The male (*left*) has his front wings elevated and is producing a series of chirps which has attracted a female, easily recognized by her simple front wings and her long egg-laying tube.

One of the most attractive passages in Gilbert White's *The Natural History of Selborne* (1789) tells of his experiences with field crickets. White observed that only the males sing and that they do so by "a brisk friction of one wing against the other." He went on to comment that "the shrilling of the field-cricket, though sharp and stridulous, yet marvellously delights some hearers, filling their minds with a train of summer ideas of everything that is rural, verdurous, and joyous."

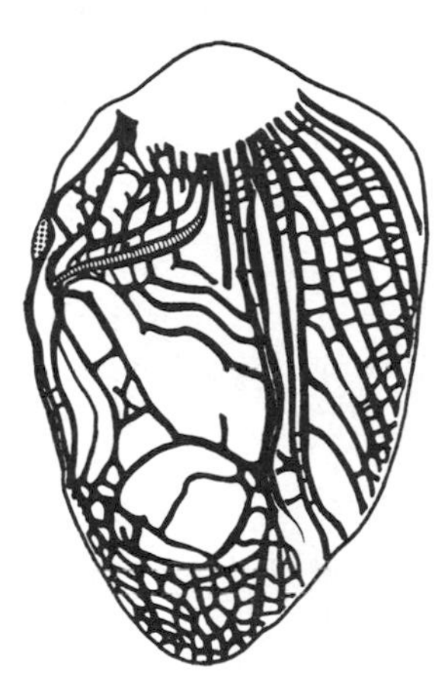

The front wing of a male cricket, from the under side. The file crosses the wing near the base, and has a series of small ridges which are rubbed against the scraper, shown here stippled. The scraper has a sharp edge on the top side of the wing, and of course it is the file and scraper of opposite wings which work together.

As White noted, the female cricket is easily told from the male by her long egg-laying tube and her simple wing structure. The front wings of the male are broadened, and their supporting struts or "veins" distorted in such a way that there are areas of unsupported membrane that are set in vibration when the cricket sings. The major sound-producing structure is formed from a major wing vein that crosses the wing near its base and bears anywhere from about eighty to three hundred minute, transverse ridges. This "file" moves against a "scraper," a hardened portion of the inner wing margin. A singing cricket raises its wings at a forty-five- to ninety-degree angle, and rubs the file of one wing over the scraper of the other, at the same time causing the wing membrane to vibrate. Since both wings of the field and house crickets have a file and scraper, it is sometimes said that they are "ambidextrous," though in fact they almost always sing with the right wing overlying the left. About 5 per cent are "left-winged," but the file of the left wing is not quite so well developed as that of the right, so these crickets produce a weak and less musical song. However, it seems to serve them pretty well. Katydids, by the way, are "left-winged," singing with the left wing overlapping the right. Since the file on the right wing of katydids is poorly developed, "right-winged" individuals are unable to sing at all. Such katydids are extremely rare and doubtless unsuccessful in attracting a mate. Incidentally, there is a genus intermediate between the crickets and katydids, Cyphoderris, in which the males sometimes use the right wing, sometimes the left—they are "switch-wingers."

The ears of crickets and katydids are present in both sexes as small disks near one of the front leg joints. As someone has said, these are the only animals that "hear with their elbows."

With their unique sound-producing organs, which we would have to classify as something between a stringed and a percussion instrument, crickets produce the purest tones known among insects. In the words of one enthusiast, crickets "alone have acquired tonality of an order closely akin to the pitch of our own formalisms of music." They have, in fact, several different songs in their repertory, but what we usually hear is the "calling song," produced loudly and for long periods by a male who is ready to mate and who is inviting females within hearing range. It is said that the song of a field cricket can sometimes be detected by man from a distance of about 150 yards, though in fact female crickets are attracted from a distance of little more than 10 yards. Some of the green meadow grasshoppers can be heard by man from a distance of over 500 yards.

The calling song of Gryllus pennsylvanicus, the northern fall field cricket of the United States, consists of a series of chirps at the rate of about four per second; but in fact each chirp consists of four pulses so close together that to a human ear they sound like a single chirp. It has been known for many years that the rate of sound production is influenced by temperature. In 1897 A. E. Dolbear, professor of physics of Tufts College, and a noted inventor as well as twice mayor of Bethany, West Virginia, wrote an article titled "The Cricket as a Thermometer," in which he presented the original formula for determining temperature from cricket chirps, now usually called "Dolbear's Law":

$$T = 50 + \frac{N - 40}{4}$$

N is the number of chirps per minute; *T* of course stands for temperature.

Professor Dolbear, who after all was a physicist, spoke of "the cricket" as if there were only one kind. Of course there are many kinds, each with its own characteristic song. Apparently he was working with the snowy tree cricket, a bush-inhabiting species that produces a series of clear, liquid notes. Unfortunately, the field cricket isn't nearly so dependable, as the rate of chirping at one temperature may vary within several chirps per second, depending on his age, mating success, and other factors we don't fully understand. Most tree crickets, and many ground-inhabiting species, produce a continuous trill in which it is quite impossible to count the number of pulses per minute by ear. Some of the green meadow grasshoppers have complex songs in which different elements in the song are influenced by temperature in different ways. So in general, unless one is sure he is listening to the snowy tree cricket, it is safer to walk to the nearest house and read the thermometer.

Professor Dolbear also made note of the fact that when many crickets are chirping at night, all members of one species in a field may chirp synchronously, "keeping time as if led by the wand of a conductor." We now know that one individual does, in fact, act as leader and the others as followers. If the leader is silenced, the followers stutter or stop, but silencing the follower, even if only one, rarely has this effect on the leader. Deafened individuals of the snowy tree cricket sing but cannot synchronize their songs with those of others. Katydids, surprisingly, are able to alternate their songs when two are singing together. That is, a solitary individual sings *ka-ty-did* about once a second (at 80 degrees), but when two are singing near each other, each sings *ka-ty-did* about half as often, and the two alternate so that the final effect is the same as if one were singing (but in stereo, so to speak). When many katydids are singing together, each tends to alternate with his closest neighbor and to be in synchrony with those alternating with his neighbor. "The result is a great, pulsing sound which fills the air for hours when there is no interrupting wind or rain." These are the words of Richard D. Alexander, of the University of Michigan Museum of Zoology, one of the foremost current authorities on the songs of insects.

Dr. Alexander also tells an interesting story about a katydid that used to sing during the daytime whenever a typewriter was in use in a neighboring room. Someone tapped the typewriter in imitation of the katydid, and the latter slowed the rate of his song phrases immediately and alternated with the typewriter. This katydid at first produced a two syllable call (*ka-ty*), as many populations do in nature, but when a three-syllable (*ka-ty-did*) song was simulated repeatedly on the typewriter, the katydid went over to a three-syllable song. By switching back and forth on the typewriter, the katydid was induced to change between three-, two-, and even one- or four-syllable songs. However, he could not be persuaded to go over to a five-syllable song, although it is said that a race of the katydid in the southern states typically has a three- to seven- (usually four- to six-) syllable song. Presumably this katydid was a Yankee.

In spite of the fact that the rate of song production is influenced by temperature and the fact that some species are influenced by the songs of their neighbors, the basic rhythm and tone are inherited and are characteristic of the species. Adult males produce the characteristic song, and females respond to it, without either having heard the song before. P. T. Haskell, of the Anti-Locust Research Centre, London, tried exposing immature female grasshoppers to a variety of sounds, but when they matured they responded only to the songs of males of their own species. It is possible (if one is very patient) to track down the common singers in one's neighborhood and to learn to associate songs with singers. In these days of "instant everything," one can of course learn the songs of crickets, katydids, and other singing insects occurring in the eastern United States by means of a record: *The Songs of Insects,* produced by Richard Alexander and by Donald J. Borror of Ohio State University. Professor Borror is one of the pioneers in the recording of insect and bird sounds.

Unfortunately, the human ear is not a very sensitive instrument for studying the songs of insects, most of which consist of very rapid pulses and some of which are so shrill that older persons cannot hear them at all. Specialists on insect sounds use a battery of instruments: microphones with parabolic reflectors to pick up the sounds, magnetic tape recorders to collect them, and oscilloscopes and audiospectrometers to analyze them in the laboratory. By the use of such devices one can readily compare the frequency, pattern, intensity, and other characteristics of the songs of a species at various temperatures, the different songs of one species, or comparable songs of different species. Nevertheless much of the early spadework in this field was done by amateur naturalists with none of this fancy equipment, and there is still plenty to be done (especially in the western United States and all over the tropics) by simply describing the songs as they sound to the human ear and associating them with the singers. Data on habitat and time of day and time of year are always valuable, and if one is truly industrious he can keep the singers in cages and study their courtship and mating in detail. One of the pleasures of entomology is that it is such a huge and undeveloped field that studies on all possible levels of sophistication can still be fruitful, and will be for a very long time.

But we have strayed into generalities from the crickets we were talking about. It is time to say a little more about their repertory and the function of the various songs they sing. We would have to disagree with Cowper that crickets are simply "full of mirth"; rather the opposite, for a male singing the calling song is a bachelor who would rather not be one. A male that has sung a long while without attracting a mate tends to increase the speed of his song, and very old, unmated males, rather than being worn out, will often scrape away furiously at the slightest provocation.

A male field cricket tends to spend the greater part of his life in one place, either a naturally occurring crevice or a burrow that he digs himself. Since males attack other males around their burrows, they tend to space themselves fairly widely, and after brief sallies they return to their own crevices. A male rendered incapable of singing does not often encounter a female, and if he does he is unable to attract her to him. On the other hand a normal male spends many hours, in fact many days, broadcasting his availability to females within hearing range. A male European field cricket was found to chirp 42,000 times in four hours and twenty minutes. Whether he found a mate at the end of that time is not recorded; I suspect that the person making the count simply went home to bed.

A female cricket moves around a good deal more than the male, and when she is ready to mate she responds to the call of the male that is loudest to her ears (therefore usually the nearest male). Her response is to stop what she is doing, then pivot to face the sound, the wide spacing of the eardrums on her two front legs permitting her to determine its direction fairly accurately. After a moment she starts toward the male, moving forward jerkily each time he sings, slowing down or stopping in between. A blinded female does just as well as a normal one, and a female proceeding with a strong tailwind does very well also, since sight and smell play no role in her behavior at this time. This can also be shown by having a male sing into a telephone some distance away and placing the receiver within range of a female, who proceeds directly to it.

Thomas J. Walker, Jr., of the University of Florida, has made a study of the response of female tree crickets to the songs of the males. Professor Walker placed virgin females in one end of an elongate cage, at the other end of which he placed a loudspeaker through which various sounds could be broadcast from tapes. The response of females to these sounds was then studied. Some were recordings of the natural songs of diverse species, while others were artificial, electronically produced sounds in various patterns and frequencies. Typically, the females responded only to the songs of males of their own species and to artificial trills having the same pulse rate; however, in species that produce chirps rather than a continuous trill, the response was principally to chirp rate rather than to the pulses within the chirp (the females responded well to artificial, pulseless chirps). It should be remembered that pulse rate varies with temperature; in fact, different species can be made to sing at the same pulse rate if the temperature of the male is properly adjusted. Can a female be made to respond to the male of the "wrong" species if the latter is heated up until his pulse rate is equal to that of the "right" species? She certainly can. The female black-horned tree cricket, for example, when kept at 70 degrees, responds well to the male black-horned tree cricket at 70 degrees, but also to the male silvery tree cricket singing at 80 degrees and the male four-spotted tree cricket singing at 89 degrees: for at these temperatures the latter two species approximate the pulse rate of the black-horned species at 70 degrees. Of course, in nature male and female would be at about the same temperature, so the species would not interbreed. If both the male and female of the black-horned tree cricket are heated to 80 degrees, the response of the female also changes, so that she moves toward the male of her own species, who of course has increased his pulse rate. If the female had a rigid, unchanging response, which did not adjust itself according to the temperature, she would end up mating with various "wrong" males, depending on the temperature!

It is the calling song, then, that tends to keep different species from interbreeding in nature. Once the female has reached the male, she touches him lightly with her antennae, and the male switches over to a less distinctive "courtship song." In the northern field cricket, the courtship song is much softer than the calling song, and when producing it the male holds his wings at a much lower angle with the body. In this species, the courtship song consists of chirps having many more pulses than in the calling song, and each terminates in a clear "tick." As he sings this song, the male backs toward the female and sways from side to side, flattening his body toward the ground. The female, if receptive, allows the male to back under her, or she may actually mount the male. As she moves forward over the back of the male, she touches his back continually with her mouthparts, and the male on his part strokes her with his antennae. Soon the female stops and lowers the tip of her abdomen; the male then raises his abdomen and attaches a sperm packet to the vulva of the female.

But that is not necessarily the end of the story. In tree crickets and in the little field cricket, the male now begins still another type of chirping, called the "postcopulatory" or "staying together" song. This stimulates the female to remain with him, and prevents her from chewing away the sperm packet before the sperms have entered her body. It also makes her available for further mating. After all, the male has worked hard to attract her, and it would not be economical to let her stray off so soon. Male tree crickets also have glands on their backs that produce a substance the female feeds upon avidly, thus keeping her available for the transfer of further sperm packets. All these lures for the female evolved long before man came on the scene, and one cricket enthusiast facetiously refers to the male's dorsal glands as "the first box of candy."

It has been noted that females of the brown bush cricket usually chew on the front wings of the males while they are perched on top of them. This has led one observer to characterize the females of this species as "a jealous set" which, having once "gained the affections of a male, devour his [wings] to keep him from calling other females. . . ." It is probably more realistic to consider this another type of adaptation for keeping the female occupied during and between copulations. The fact is that in most populations of this species the male does not produce a calling song but hunts out the female and follows her about until she succumbs to the temptation to nibble his wings, which he does not use for singing anyway.

Even in the common field crickets, which lack dorsal glands and postcopulatory songs, the males often succeed in keeping the females with them for several matings. They do this by repeatedly touching her antennae and by keeping other males away. Richard Alexander tried putting two males and one female in a cage together, and he found that the more dominant of the two males "in every case completely monopolized the female for hours at a time, copulating with her at intervals of an hour or less. . . ." When Dr. Alexander temporarily penned the dominant male in a corner of the cage and let the other male court and mate with the female, this second male continued to hold the attention of the female, fighting off the first male whenever he approached.

This leads us to still another type of song, the so-called "rivalry song," or "aggressive chirps," produced by one male when repelling another. These are short, sharp sounds of rather irregular occurrence that tend to "warn" other males or to accompany or follow actual combat.

Some crickets have still other sounds in their repertory. Burrowing crickets, for example, often produce "recognition chirps," which apparently help them to keep in touch with one another in the soil. These are well developed in mole crickets, which incidentally are among the very few crickets in which the females also produce sounds. The low, liquid notes of the mole cricket are to my mind one of the most delectable of insect sounds, perhaps because they remind me of nights camped beneath live oaks strung with Spanish moss along Florida streams.

Among the grasshoppers still other kinds of songs occur. Sometimes, for example, the female produces an "agreement song," considered an indication that she is willing to copulate. In a few grasshoppers the male produces a sharp noise just before he leaps onto the female, a sound that some imaginative entomologist has termed the "shout of triumph." (Grasshoppers, just to be different from crickets, mate with the male on top of the female.)

So in fact sounds play many roles in the lives of crickets and grasshoppers. A cricket without his fiddle is as badly off as a blind dragonfly or a roach that cannot smell. Yet there are a few mute crickets. Some occur in caves, and have developed enormous dorsal glands that presumably compensate for the loss of their sound-producing organs. The tiny cricket Myrmecophila, which lives as a guest in ant nests, is also silent. Both types of crickets lost their wings when they became adapted to their special habitats, and with their wings went their voices.

But it is the less secretive kinds of crickets that we are usually aware of, and these more by song than by sight. Their calling songs may remind us of all that is "rural, verdurous, and joyous," as Gilbert White remarked, but to the crickets they are highly functional. It is a fact that no two species of crickets, katydids, or grasshoppers occurring at the same time and place have identical songs, for it is by the song that the female locates the male of her own species. However, crickets occurring in different geographic areas or maturing at different seasons sometimes have identical or nearly identical songs, for here there is no danger of confusion.

Since many crickets look pretty much alike to the human eye, it is not surprising that entomologists, too, found that the best way to distinguish species was by their song. At one time it was supposed that all the field crickets in North America belonged to one species. Then, beginning in the 1930's, the late B. B. Fulton, of North Carolina State College, one of the finest naturalists of his generation, found that in his state there were four different kinds of field crickets, distinguishable by their songs and habitats. He called these the triller, the woods cricket, the mountain cricket, and the beach cricket. He tried crossing males of one kind with females of another, but in fifty trials he failed to obtain any offspring at all. But in twenty-five cases in which a cricket was paired with its own kind, only two failed to produce offspring.

Clearly there were four species of field crickets in North Carolina, even though museum workers could only classify their dead specimens in one. Later, Richard Alexander distinguished two more species in the eastern states, one of which he appropriately named Gryllus fultoni, the southern wood cricket. Analysis of the calling songs with an audiospectrometer revealed that all but two of them showed differences in pulse rate and pattern. These two were the so-called northern spring field cricket and the northern fall field cricket, which, because they mature at different seasons, never meet in nature anyway. Here is Dr. Alexander's comparison of the songs of the six field crickets of eastern United States:

	Triller cricket	Beach cricket	No. wood cricket	So. wood cricket	No. spring field cricket	No. fall field cricket
Type of song	trill	chirp	chirp	chirp	chirp	chirp
Pulses per second (85°)	60	17–19	30–33	43–50	24–29	24–29
Chirps per minute (85°)	—	100–120	180–200	300–360	150–240	150–240
Pulses per chirp	—	4–6	2–4	2–4	3–5	3–7

Dr. Alexander verified and extended Fulton's experiments on interbreeding. Sometimes females would mate with males of another species in laboratory cages, but in only a very few cases were hybrids produced. These few hybrids had songs intermediate between those of the parent species, and in nature the males would very likely call in vain. Alexander did find one naturally occurring male hybrid between the southern wood cricket and the northern fall field cricket. This individual had "the pulse rate of one parent, the chirp length of the other, and a chirp rate intermediate between the two." Incidentally, when the northern spring and fall field crickets are made to mature at the same time in the laboratory, these two species, with their identical songs, fail to produce any hybrids at all. Obviously the songs of crickets are inherited, but just in case there is a breakdown in auditory or seasonal isolation, there are still other mechanisms that we do not fully understand which prevent these species from producing offspring (except very rarely).

Needless to say, it came as a considerable shock to find that "the" field cricket of the eastern states was really six species. But in fact several other types of crickets were found to have "hidden" species, that is, species we are hard put to tell apart by sight but which still maintain themselves as distinct, noninterbreeding populations in nature. There is reason to believe that "hidden" species are common in many groups of insects, and it may be vitally important to be able to identify them (for example, some species of Anopheles mosquitoes transmit malaria, while others, distinguishable only by shape and color of the eggs and by certain behavioral features, do not transmit it). We might visualize a research museum of the future in which every species would be represented not only by dead specimens of the adults but also by representatives of all life stages as well as by motion pictures of characteristic behavior patterns, tape recordings of songs, analyses of glandular secretions, and so forth. Such a museum would have to be large and well financed, but in fact we could build, staff, and operate many such museums with the 22 billion dollars it is costing us to place a man in that most sterile of all landscapes, the moon.

Before leaving the cricket to his fiddling, we should say a bit more about the behavior of the males toward one another. Here again we may dip into the research of that ubiquitous cricket enthusiast Richard Alexander. Dr. Alexander collected immature male crickets, and reared several of them together in a terrarium. As immatures, they were primarily interested in eating, but as each molted to the adult stage he became aggressive toward other adults in the cage. When two adult crickets approached, they lashed each other vigorously with their antennae. If one or the other did not withdraw, they reared up and grappled with their front legs, at the same time raising their wings and producing aggressive chirps. If neither retreated at this point, they began a real wrestling match with their legs and jaws, and the weaker was sometimes flipped over and rarely even lost a leg or an antenna. However, only a few fights proceeded this far; in most cases one of the crickets succeeded in dominating the other by sheer tenacity.

In Dr. Alexander's cages the males soon set up a dominance hierarchy, or "peck order" (a term that is a holdover from the days when such hierarchies were known primarily in chickens). That is, the "top" male usually won all his fights; the second in order won most fights except those with the number-one male; and so forth. Encounters between males ranking close together in the hierarchy were often intense, while those between individuals far different in position were usually broken off at an early stage. Males attained their peak dominance about twelve days after reaching adulthood, and larger males usually tended to dominate smaller ones. It was found that certain conditions can cause a temporary improvement in a cricket's position in the peck order. For example, if a cricket is held in isolation for a period and then introduced into a group, he proceeds to win most of his fights, at least for a few hours. Also, a male that has just copulated with a female becomes very aggressive, and is able to dominate all the other males for a short while. Most interesting of all, a male that has established a "home," that is, a burrow or niche in which he remains most of the time, is unusually aggressive toward crickets he encounters when he occasionally sallies forth. When all the crickets in a large terrarium establish their own territories, they tend to keep each other "at bay" by aggressive chirping and occasional combat, so that the amount of fighting is much reduced. This, of course, is what happens in nature, and the fact that the males space themselves widely and remain in their territories most of the time is intimately tied in with reproductive behavior, as we have seen.

Now that we know so much about the private lives of crickets, and can rear them and manipulate them so easily in the laboratory, it is not surprising that some investigators have decided to see if they can find out "what makes them tick." We know that much of their behavior, like that of other insects, is inherited and is programmed inside their bodies. It is going a little far to compare insects to mechanical toys, as some persons have done; it is not quite so simple as that. Nevertheless, when we know that one species chirps 100 to 120 times a minute and a closely related species 300 to 360 times a minute (at the same temperature), it is natural to ask what mechanisms control chirping and other instinctive activities of crickets. It will be a long while before we can answer a question like that fully, but Franz Huber, of the University of Köln, Germany, has made a beginning.

The cricket is a wonderful animal for "tinkering." He can not only be reared in large numbers, but he is reasonably large and not covered with long hairs or scales like many insects. The cricket can be anesthetized, and a "handle" glued to the shield on the back just behind the head. One can then cut a hole through his hard body wall almost anywhere and insert probes or wires and see what happens. One of the simplest operations is to cut a small window in the front of the head and to damage certain parts of the brain with a needle. Severe injury to many parts of the brain renders the cricket mute; however, after damage only to the "caps" of the so-called "mushroom bodies" of the brain, males sing continuously until completely exhausted. While other parts of the brain evidently play a role in the production of song, this particular region appears important in stopping it. The mushroom bodies, by the way, received their name because they have a "stem" and a "cap" suggestive of a mushroom; if this name sounds a bit too silly, one may call them "corpora pedunculata." There are two of them toward the extreme front of the insect brain, and by whatever name one calls them, they are undoubtedly major coordinating centers in the insect body.

Professor Huber extended these experiments further by inserting very delicate tungsten wires into the brain and by stimulating specific points electrically. It was not possible to select each point precisely, since the cricket brain is a small but very complex structure. What he did was to make an insertion of the wire, send an electric impulse through it, and record the resulting behavior. Having done this, he would kill the cricket and, using refined techniques, section the brain and study it under high magnification. In this way he could determine the exact point in the brain reached by his wire. By doing this many times he was finally able to "map" the brain, that is, to determine what particular areas in the brain were associated with particular behavior patterns.

Huber confirmed the fact that the caps of the mushroom bodies inhibit singing. Stimulation of certain areas in the stems, however, produces a song resembling the calling song characteristic of that species. Very slightly deeper probing of the brain, along the inner sides of the mushroom bodies, results in aggressive chirping and antennal slashing and other movements characteristic of fighting behavior. Finally, stimulation of the interior part of the brain, in and around the so-called central body, causes the crickets to raise their wings and rub them together quickly and irregularly, producing a whirring sound similar to that made just as the cricket prepares to fly.

Similar experiments were also tried on the nerve mass located in the middle section of the thorax and supplying nerves to the wings and middle legs. When this nerve mass was cut off from the brain and stimulated electrically, chirping resulted, but it followed no set pattern, and in fact the experimenter could set his own pattern by altering the electrical impulses. One can conclude, then, that this thoracic nerve center does control the muscles of the wings, which in turn move the wings and result in sound production. However, the brain is involved in telling the thoracic center when to start and when to stop. This does not

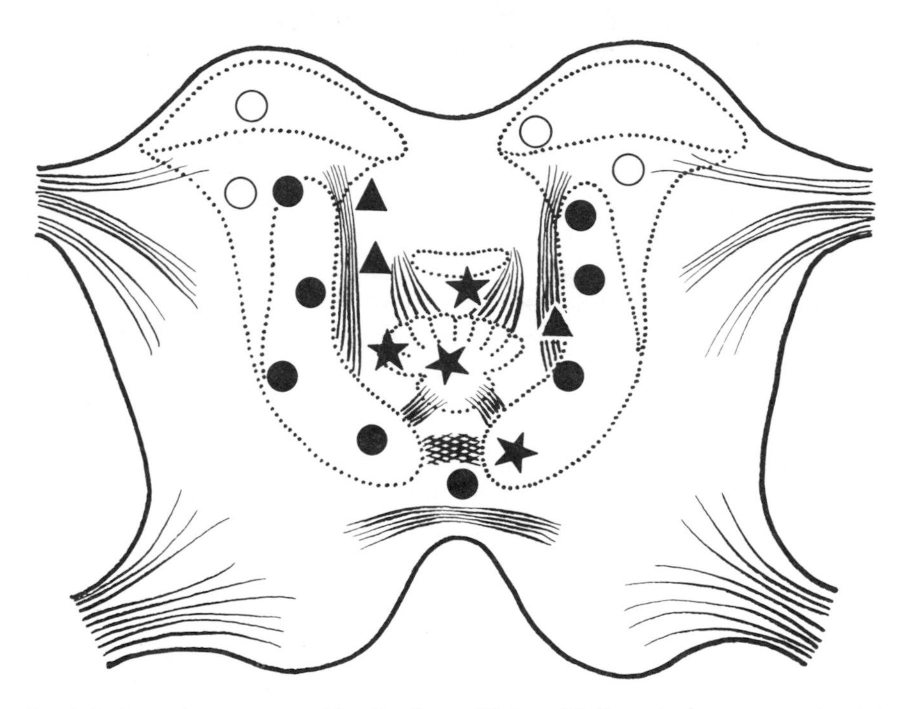

A cricket's brain, as mapped by Professor Huber. Hollow circles represent points at which singing was inhibited. Solid circles represent points at which a song resembling the calling song was elicited. When points represented by triangles were probed, aggressive chirps and antennal slashing occurred, while probes at the points represented by stars resulted in wing-whirring. The large, mushroom-shaped structures are the so-called mushroom bodies; the small, paw-shaped structure in the center is the central body. Major nerve tracts are shown by dark lines. (After Franz Huber, somewhat simplified.)

necessarily mean that the pacemaker resides in the brain, however. Aggressive chirps are triggered by antennal stimulation, and so are subdued calling songs similar to those Huber obtained from the stems of the mushroom bodies (just as the male contacts a female and prepares to switch to the courtship song). Since nerves from the antennae pass through these parts of the brain, Huber may merely have been simulating certain types of antennal stimulation, and the impulses may have been carried to a pacemaker elsewhere. At any rate it is a start toward an understanding of the situation. It seems probable that the genetic differences between species actually involve slight differences in nervous connections and rates of firing of nerves within specific nerve centers. The mechanism that dictates that the female respond only to the song of the male of her own species must also lie in such centers. Professor Huber concludes a recent review of this field with a remark we have heard many times before, but which seems especially appropriate when discussing matters relating to the nervous system: "Many questions are still unsolved."

It is a pity that in our Western world the study of crickets has been left to a few fortunate specialists. In our busy world, most of us haven't even the time to pause and listen to them; they may sing all about us, but we are so preoccupied with our own affairs that we do not hear them. In China, crickets have long been revered as vocalists and as pugilists. Berthold Laufer, former curator of anthropology at the Field Museum in Chicago, and a noted authority on Oriental culture, has written a fascinating essay, *Insect-Musicians and Cricket Champions of China.* According to Dr. Laufer, from times of antiquity down through the sixth century A.D., the Chinese merely appreciated cricket songs, but eventually they learned to keep them in cages in their homes so as to enjoy them more fully, and during the Sung dynasty (960–1278 A.D.) they developed the sport of cricket fighting. An account dating from 742 to 756 notes:

"Whenever the autumnal season arrives, the ladies of the palace catch crickets in small golden cages. These with the cricket enclosed in them they place near their pillows, and during the night hearken to the voices of the insects. This custom was imitated by all people."

The "people," of course, could not afford golden cages, so they built them of bamboo. Some were designed for wearing suspended from the belt, so that cricket songs could be enjoyed at all times of the day. Special traps for catching crickets were designed, as well as various devices for feeding and handling crickets. During the summer, the crickets were reared in bowls with flat lids, often elaborately constructed by potters who specialized in cricket cages. In the winter they were transferred to gourds with perforated covers. These gourds were grown in earthern molds of various shapes so that they assumed certain prescribed forms, and were decorated with intricate designs. The covers were sometimes made of sandlewood, jade, or ivory, carved in designs of dragons or flowers. Some of these cricket houses are considered priceless works of art.

The Chinese have developed an extensive literature on the care and feeding of crickets. The *Tsu chi king,* or *Book of Crickets,* written in the thirteenth century, told of the several kinds of crickets occurring in China, and gave detailed instructions on how to rear them. Different diets were recommended for different species, and cricket fanciers often developed their own special diets for their pets. Common foods consisted of cucumbers, lettuce, honey, chopped fish, and even beans and chestnuts previously masticated by their masters. Fighting crickets received a special diet containing lotus seeds and mosquitoes, and just before the fight they received a special tonic made from the roots of certain flowers.

Down through the centuries the Chinese have developed criteria for selecting crickets for their vocal powers or their fighting abilities. "Cricket-fights in China," Laufer says, "have developed into a veritable passion. Bets are concluded, and large sums are wagered on the prospective champions. . . . Choice champions fetch prices up to $100, the value of a good horse in China. . . . In southern China, a cricket which has won many victories is honored with the title 'conquering or victorious cricket' (*shou lip*); on its death it is placed in a small silver coffin, and is solemnly buried." Laufer describes a cricket fight as follows:

"The tournaments take place in an open space, on a public square, or in a special house termed Autumn Amusements. There are heavy-weight, middle and light-weight champions. The wranglers are always matched on equal terms according to size, weight, and color. . . . As a rule, the two adversaries facing each other will first endeavor to flee, but the thick walls of the bowl or jar are set up as an invincible barrier to this attempt at desertion. Now the referee . . . intercedes, announcing the contestants and reciting the history of their past performances, and spurs the two parties on to combat. For this purpose he avails himself of the tickler [a special instrument consisting of rat whiskers inserted in a bone or ivory handle, providing an obvious substitute for antennae lashing] and first stirs their heads and the ends of their tails. . . . The two opponents thus excited stretch out their antennae . . . and jump at each other's heads. . . . One of the belligerents will soon lose one of its [antennae], while the other may retort by tearing off one of the enemy's legs. The two combatants become more and more exasperated and fight each other mercilessly. The struggle usually ends in the death of one of them. . . ."

Just as the Japanese have vernacular names for all their many kinds of dragonflies, so the Chinese have one or more names for each species of cricket. In parts of Africa, too, crickets are cherished, and their songs are believed to have magic powers. It is said that when the island of Jamaica was discovered, many of the Indians were seen to be carrying baskets of crickets. It is a pity that Western man has become too busy, too preoccupied with his own noisy world, to listen to the cricket. True, we know that the cricket is not quite the cheerful troubadour the romantics supposed him to be. Indeed, at times he does seem to inhabit a world "more insensate, more atrocious, more infernal than ours," to repeat Maeterlinck's words. Nevertheless he illustrates admirably some basic animal traits not wholly alien to man. His love life may not be quite like ours, but he is equally preoccupied with it, and after all he invented the serenade long before we did. No one who has ever dwelt in the business world—or for that matter the academic world—can doubt man's tendency to establish a peck order. Like the cricket, man is territorial and aggressive. Robert Ardrey, in his book *The Territorial Imperative,* maintains that modern war is the result of territorial instinct upon which is imposed a technology responsible for death-producing devices far more efficient and more impersonal than could have been imagined just a few years ago. That man could understand or improve his behavior by developing a closer rapport with the cricket is, however, doubtful; after all, he has never learned very much from his own history.

I do believe that an intimacy with the world of crickets and their kind can be salutary—not for what they are likely to teach us about ourselves, but because they remind us, if we will let them, that there are other voices, other rhythms, other strivings and fulfillments than ours. A symbol of our times is the transistor radio, capable at any time and place of blaring forth the rhythms that man has evolved down through the centuries. There is other music than this (thank God), and it is good to slacken our pace now and then and listen to patterns of sound that fell on meadow and forest millions of years before there were men to hear them; patterns that in their simplicity and repetitiveness suggest the heartbeat of life itself or the ticking of some cosmic clock. What was it that Nathaniel Hawthorne said of the tree cricket? "If moonlight could be heard, it would sound like that."

While the chirping of a single cricket is not likely to hold our attention for very long, the chorusing of myriads of crickets can be bewilderingly complex; furthermore, the chorus changes, depending on the locality and the time of year. Some years ago the late Robert Evans Snodgrass, one of the most distinguished entomologists of his day, described the seasonal sequence of insect song near Washington, D.C. "The Season's Program," he called it:

"There is something peculiarly quiet about the evenings of spring and early summer. . . . But, on one of those first warm evenings toward the end of May, when the air is motionless though not yet sultry in southern Maryland, as the sun's fiery tints on the fleecy lining of the sky give way to the paler tones of moonlight, there comes from somewhere on the lawn the first heralding of that troupe of insect choristers that later will make the night air ring from dusk to dawn with the strident music of their serenades. The announcement is a cheerful *chirp, chirp, chirp,* strong and clear but vibratory, the unmistakable notes of Gryllus, the common black cricket. . . .

"The season advances, the hot nights of July arrive, thunderstorms come and go, each leaving in its wake that oppressive evening stillness reeking with moisture. On such a night, perhaps, toward the middle of July, as you may be meditating on nature's mightier forces, suddenly, from a near-by tree or shrub, you hear a voice, or the semblance of one, which says: *treat, treat, treat, treat,* regular repetitions of one note 140 times a minute. Over there it is echoed by another, and over yonder by another. The next night still more join in, and soon the very atmosphere vibrates to that monotonously measured beat. This is the music of the snowy tree cricket, first of the summer chorus to arrive upon the stage.

"A few nights later another sound cuts across the rhythmic concert, a longer, purring note, sad and melancholy in tone as if from some complaining spirit of the night, a soft *burr-r-r,* prolonged about two seconds and repeated at intervals of equal length. This is the song of the narrow-winged tree cricket, a cousin of the snowy. Representatives of his species likewise come on in greater numbers every evening till soon the nightly chorus is a blend of *treats* and *burrs.*

"Then to add to the confusion, several other bands arrive that strike up long unbroken trills. . . . The chorus of the snowy and the narrow-winged now falls into the background and individual voices are lost as the entire concert becomes a ringing and shrilling arising at twilight, increasing with darkness, so invisibly linked with the oncoming shadows that it seems almost to be an emanation from the night itself. [Soon various katydids join it, then the fall field cricket, along with] another smaller cricket of the turf, a very little fellow called Nemobius, with a very little voice, and one so delicate that you must bend low to catch his elfin notes, a silvery, twittering trill, rising like the music of some unseen pygmy from the grass.

"Shortly after the advent of the second Gryllus band a new note breaks out, a loud, piping chirp inflected upward at the end, a sound easily mistaken for that of some little tree toad. The notes are hard to locate precisely, they seem to come from here, from there, from over yonder, and from back here again, but always singly. . . . They are the song of another cricket, the jumping bush cricket . . . the last performer to appear on the nocturnal program. No other new notes after his are to be heard, though the voices of the others will change. As the cold of fall increases, the clear *treat, treat, treat* of the snowy cricket becomes a broken rattle, the sonorous purr of the narrow-winged changes to a long hoarse rasp, the notes of the trillers become weak and subdued, the cheerful chirps of Gryllus feeble and shaky. . . . At last the killing frosts arrive, insect voices are stilled, and the season's concert series ends."

Let me conclude this chapter with a quote from another cricket admirer who, like Snodgrass, was for many years associated with the United States Department of Agriculture—H. A. Allard:

"The study of the musical moods and behaviors of the insect-instrumentalists of the earth is no small matter. . . . This specialized interest may seem very remote from our practical affairs of a mechanized age, but the basis of all pure science is healthful curiosity; and truth, whatever its nature or revelations, is worthy of record in the archives of men. The singing cricket on my hearth may not add to the contents of my pocketbook, but at times it affords an indefinable inspiration and poetry that reveals new beauties of living, of expression, and of association in the universe. I am better, broader, wiser, happier for having heard the crickets and katydids, for somehow there are points of kinship in our lives, even though our magnitudes and roles of living seem so far apart."

In Defense of Magic: The Story of Fireflies

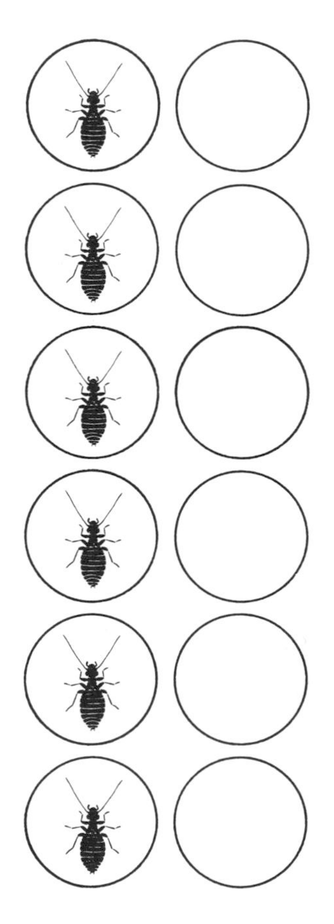

If magic be defined as something "produced by secret forces in nature," and "secret" in turn defined as something "revealed to none or to few" (and these are legitimate definitions), then magic is not likely to be diminished by all the science we can muster. Research may provide us with answers, but these answers forever lead to new and more profound questions; and as our knowledge of the world grows more and more vast, most phenomena that can be said to be "revealed" will of necessity be revealed to fewer and fewer of us. True, the universe is orderly (more or less) and finite (probably), so it is possible to imagine a time when all knowledge has been stored in a system of computers from which any fact or combination of facts can be elicited by pushing a button. As I say, it is possible to imagine this. But just recently I have been reading several articles on animal behavior written between 1900 and 1920. Their language seemed a bit naïve by modern standards; but I wondered if we had really advanced very much in this field in half a century. Recently, too, I have been studying a group of small wasps known as the

Bethylidae, a group known to only a few persons and in which probably more than half the species are, in fact, unknown to anyone—and we don't really "know" any of the several hundred "known" species in any real sense, although a few of them are reported to have rather remarkable behavior patterns. True, some of my colleagues are deciphering the genetic code. But how much can they tell us about the production of a bethylid wasp from a particular series of chemical messages? And why, of all possible concatenations of atoms, are there wasps at all?

Magic in the sense of something "inciting wonder" is also here to stay; or if it is not, man will have been vastly diminished by its loss. One need not be standing silent on a peak in Darien. There is magic, indeed, in the crash of surf on an unknown shore; but so there is in a mud puddle. There is a powerful magic in a crash of thunder, even more powerful in a nuclear explosion; but there is a very special magic in a child's kite or in the call of a gull and all that it evokes. Mark that leaf blown before the wind: it is important; no matter how sophisticated or blasé we become, that moment, this experience, is all the treasure we shall reap in our few moments of identity.

What can rival a twilit meadow rich with the essence of June and spangled with fireflies? Here is magic, indeed, and the joy of pursuing through grass just touched with early dew a light now here, now there, now gone. Or of collecting several in a bottle and taking them indoors for illumination; or of tying one lightly with a thread to one's clothing, as natives of some tropical countries are reported to do at fiesta time. As children, we used to call them lightning bugs; and wingless kinds that emit a steady light from the ground are called glowworms in English-speaking countries wherever they occur. In fact fireflies are neither flies nor bugs nor worms, but soft-bodied beetles called Lampyridae, a name based on an old Greek word that also evolved into our word "lamp."

Some of our commonest Lampyridae, curiously, give no light at all; these are day-flying beetles that one often finds on tree trunks, looking very much like ordinary fireflies but lacking the whitish "lamps" in their tails. The common European glowworm is a wingless female that produces a steady light, while the male of the same species is winged and not luminescent. Most fireflies of eastern North America are winged, and produce a flashing light in both sexes. The larvae (and even the eggs!) of many fireflies also glow. This seems strange when we consider that the lights of fireflies are used by the adults to find the opposite sex of their own species in the dark. What function does luminescence serve in the eggs and larvae? One might assume that the immature stages simply "can't help glowing," since the rudiments of the light organs are developing within them. But the fact is that the larval and adult organs are of quite different nature, and if the larval light-producing cells are carefully excised, the adult will still develop normal light organs.

Luminescence probably first arose as a dim and diffuse product of certain normal body processes, for many substances oxidized slowly in the dark produce a glow, and a dim luminescence occurs in many simple organisms (especially in the sea). Natural light is known to occur in certain bacteria, fungi, one-celled animals, sponges, jellyfish-like animals, corals, marine worms, clams, snails, squids, arthropods, and of course a variety of deep-sea fishes—but never among the reptiles, birds, or mammals. It is possible that the earliest organisms on earth lived in an atmosphere devoid of oxygen. When oxygen first appeared—from the effects of sunlight on water vapor or from photosynthesis by primitive plants—it may have been toxic to these organisms. Luminescence may have developed as a system of getting rid of oxygen by burning it off as a "cold light." Later on, when plants and animals evolved that took advantage of oxygen to run their own body machinery, luminescence was preserved in a wide variety of organisms simply as a hangover from these ancient times. At least such is the belief of William McElroy and Howard Seliger, of Johns Hopkins University, our current leading authorities in this field. Their theory is supported by the fact that in many simple organisms luminescence seems to serve no function, and in fact in some cases a single species exists in both luminous and nonluminous forms, both apparently successful. They also point out that luminescence requires oxygen in only very low concentrations, as it must have once occurred on earth. Certain bacteria, for example, produce light when the oxygen concentration is as low as one part in 100 million.

Obviously, some of the more complex animals—fish and insects, for instance—have elaborated this primitive light-producing capacity into specialized organs serving important functions in their lives. Adult fireflies possess the most complex light organs known, and these organs are still far from fully understood. Despite the intensity of the light they produce, the amount of heat is negligible. Only in very recent years has man developed chemical light-producing systems that rival that of the firefly in efficiency.

E. Newton Harvey, of Princeton University, has written a fascinating account of the history of human knowledge of luminescence. According to Professor Harvey, the firefly is not mentioned in the Bible, the Talmud, or the Koran, probably because fireflies are absent or uncommon in the arid regions of the Near East. However, the Chinese *Book of Odes*, dating from 1500 to 1000 B.C., speaks of the "fitful light of glowworms," and there are many accounts of fireflies in ancient writings of the Far East. The Japanese believed fireflies to be transformed from decaying grasses, while glowworms were said to arise from bamboo roots. In

Japan, firefly collecting was popular in early times, and there is said to have been a firefly festival each year near Kyoto.

Aristotle was familiar with fireflies, and was apparently aware that some glowworms are the larvae of winged fireflies. The Roman encyclopedist Pliny believed that fireflies turned their lights off and on by opening and closing their wings, a statement repeated again and again down through the Middle Ages, along with a great deal of other misinformation, including tales of luminous birds. Thomas Mouffet (1553–1604) was aware that the British glowworms were females and that the males were nonluminous flying insects. Like many persons of his time, Mouffet was most interested in the medical uses of plants and animals. Fireflies, says Mouffet, "being drank in wine make the use of lust not only irksome but loathsome. . . . It were worthily wisht therefore that the unclean sort of Letchers were with the frequent taking of these in Potion disabled, who spare neither wife, widow nor maid, but defile themselves with lust not fit to be mentioned."

The scientific study of insects is sometimes said to have begun with the publication of Ulysses Aldrovandi's *De Animalibus Insectis* in Bologna in 1602. Aldrovandi included a fairly accurate sketch of a glowworm, as well as the interesting hypothesis that fireflies use their lights to find their way about at night. A few years later Francis Bacon expressed curiosity that these insects were able to produce light without heat, but the times were scarcely ripe for a solution to this problem. The first important book on animal lights was written by the Danish physician Thomas Bartholin in 1647. Bartholin's own experiments failed when his glowworms escaped from the cage, but he discussed the unpublished work of Vintimillia, an Italian who observed the mating of fireflies in glass jars. Vintimillia was well aware that the flashes serve to attract the sexes, and he was the first to note that the eggs are luminous. During the eighteenth and nineteenth centuries a great many persons turned their attention to the life histories and luminescent properties of fireflies, including such notables as Michael Faraday and Louis Pasteur. We nevertheless still have a long way to go; one can imagine a scientist of the year 2068 looking back to our time with somewhat the same amusement we now look back on Thomas Mouffet, although considering our increasing unmindfulness of the past, it is equally possible to imagine that in 2068 men will not look back at all. Or that there will not be men at all.

Perhaps the most notable contribution to an understanding of the light of fireflies was made in 1885 by the French physiologist Raphael Dubois. Dubois removed a light organ of the beetle Pyrophorus, ground it up in water, and left it until the light went out of its own accord. He then removed another organ and ground it in boiling water for a short time, so that its light, too, was extinguished. Then he performed a neat bit of magic: when the two extracts were placed together, the light reappeared. He deduced that two substances were required to produce light, one of which was inactivated by heat. He called these two substances luciferin and luciferase (after Lucifer, who among other more devilish traits was the bearer of light). Dubois also learned how to obtain luminous bacteria from the skins of dead fish and squids on the seashore. The bacteria could be transferred to culture plates, where they produced large colonies that glowed with a blue-green light. At the International Exposition in Paris in 1900, Dubois created a sensation by lighting a small room with flasks containing suspensions of these luminous bacteria.

A good deal more has now been learned about the production of animal light, and luciferin and luciferase have been obtained in purified crystalline form. McElroy and his colleagues at Johns Hopkins have synthesized luciferin. We now know that something more is needed: adenosine triphosphate (ATP). ATP may be less familiar to most persons than DDT or the CIA, but it happens to be even more important to us, providing as it does the energy for muscle contraction in animal bodies, including our own. In the light organ of the firefly, ATP energizes not muscles but the luciferin-luciferase system, the energy appearing not as mechanical work but as light. It has recently been proposed that luciferin and luciferase be employed in automated laboratories sent to Mars or other planets. The idea is that a scoop would pick up soil from the surface and mix it with water, oxygen, luciferin and luciferase. Then if a glow were televised back to earth, we would know that ATP, the fifth requirement for firefly-light production, occurs there. The presence of ATP would mean, in turn, the existence of some kind of animal life in that alien soil. Thoughts such as these emphasize the need for caution when labeling the study of fireflies (or anything else) "useless" or "idle curiosity."

In the living insect, an additional element is needed to account for the working of the system: some sort of nervous control. It was discovered long ago that cutting off the firefly's head caused the flashing to cease, although in some cases the light organ glows dimly for a long time. Later it was found that by electrical stimulation of the severed nerve cord one can produce experimental flashing. It is believed that nervous control is centered in the brain (much like the control of chirping in the cricket); impulses then travel to the light organs via the nerve cord and via delicate nerves that closely parallel the minute tubes that carry air to the light cells. We still do not know exactly how the flashing is triggered. Some have claimed that the nerves control the supply of oxygen to the light cells, but recent work suggests that the oxygen supply may

The flashing of a male firefly. The light organs are located beneath the abdomen and provide a unique means of communication in the dark.

be constant and that the series of chemical reactions resulting in a light flash may be initiated by the synaptic fluid of the nerve endings. That there are chemical intermediaries between nerve and light organ is suggested by the fact that a nervous shock provided directly to the light organ produces a very quick flash, whereas a stimulation to the nerves always involves a longer delay than nerve conduction itself would require. These are profound matters that we understand only poorly. Indeed, we still have much to learn as to how the chemical energy supplied by ATP is converted into the mechanical energy of ordinary muscle contraction.

The light organs of fireflies are complex structures, and recent studies using the electron microscope show them to be even more complex than once supposed. Each is composed of three layers: an outer "window," simply a transparent portion of the body wall; the light organ proper; and an inner layer of opaque, whitish cells filled with granules of uric acid, the so-called "reflector." The light organ proper contains large, slablike light cells, each of them filled with large granules and much smaller, dark granules, the latter tending to be concentrated around the numerous air tubes and nerves penetrating the light organ. These smaller granules were once assumed by some persons to be luminous bacteria, but we now know that they are mitochondria, the source of ATP and therefore of the energy of light production. The much larger granules that fill most of the light cells are still of unknown function; perhaps they serve as the source of luciferin.

Actually, the light organs vary a good deal in different kinds of fireflies. We also know that the color of the light varies in different species and that this is a real difference in light color and not the result of a tinting or filtering effect of the window. Generally speaking, the light is yellowish, but it may have a greenish, bluish, or orange hue.

McElroy has found that the color of the light produced by luciferin can be changed by altering the alkalinity of the solution, less alkalinity producing a shift toward the red end of the spectrum. Present evidence suggests that various species of fireflies have slightly different luciferase molecules, which cause the production of light of slightly different wavelengths. In the genus Pyrophorus (not really a true firefly, but a click beetle) there are two greenish lights just behind the head and an orange light on the abdomen. I well remember my first acquaintance with Pyrophorus. We were camped out near the ruins of Xochicalco, in Morelos, Mexico, when a disturbance caused me to peer out into the darkness: only to find that we were surrounded by pairs of glowing green eyes. The ghosts of Toltec warriors a few yards away? No, it proved to be a host of Pyrophorus in the bushes only a foot or two away. The story is told that when Sir Robert Dudley and Sir James Cavendish first landed in Cuba, they saw great numbers of lights moving about in the woods. Supposing them to be Spaniards with torches, ready to advance upon them, the British withdrew to their ships and went on to settle Jamaica. In this manner Pyrophorus may be said to have changed the course of history.

The South American "railroad worm" is an elongate glowworm having eleven greenish lights down each side of the body and two red lights on the head. These lights are quite brilliant, and when the insect is moving along the ground it looks like nothing so much as a fully lighted railroad train. The North American railroad worm is larger but lacks the red lights on the head. Both of these insects are quite rare.

We now know that there are not only differences in the nature, shape, and position of the light organs and in the color of the light of fireflies but also (and most particularly) in the behavior patterns of the male and female during courtship and mating. The males of the European glowworm fly toward a light only if it is of the shape, color, and intensity of that of the female of that species. In our common North American species, the females often rest on the ground or vegetation, and flash only in response to the flashes of the males. In one of the best-studied forms, Photinus pyralis, the male flies near the ground in a strongly undulating pattern; he approaches the bottom of one of these undulations every six seconds, and as he does so he makes a half-second flash, at the same time rising and thus describing a "J" of yellow-green light. If he passes within a few feet of a female, the latter responds with a half-second flash of her own, but only after an interval of about two seconds (with only slight variation). This interval is an all-important signal to the male; we know this because the male will respond to various flashes, including even that of a flashlight, but *only* when these occur about two seconds after his own flash. If the female flashes at the proper interval, he flies toward her and flashes again, whereupon the female again responds in two seconds. This may be repeated several times until the male reaches the female and mates with her. There is no evidence that sound or smell play any role in firefly mating.

The larger fireflies of eastern North America belong mostly to the genus Photuris, a confusing group in which the males show much variation in flash pattern but hardly any differences in structure or body color. For many years this problem bothered H. S. Barber, beetle specialist of the United States National Museum (not to be confused with H. G. Barber, a specialist on true bugs who worked at the National Museum at the same time—the two were "beetle Barber" and "bug Barber" to their colleagues). The results of H. S. Barber's study were not published until a year after his death in 1950. Barber found that in the Potomac Valley he could detect a woodland species with a short greenish-white flash once a second; a stream-side species with a slightly slower, faintly orange flash; a species occurring in alder groves and poising almost motionless, its light beginning dimly and growing steadily in brilliance before stopping abruptly, only to reappear at a different point several seconds later; and so forth. Eventually Barber recognized eighteen species of Photuris, mainly on the basis of the flashes of the males; ten of these he had to name as new, since they had not previously been recognized. Needless to say, this did not endear him to museum workers, who could not very well sort their dead beetles on the basis of their flashes. But as Barber said:

"Taxonomy from old mummies which fill collections is a misguided concept. It leads to the misidentification of rotten old samples in collections. How these poor fireflies would resent being placed in such diverse company—among specimens of enemy species—if they were alive and intelligent! What contempt they would feel for the 'damned taxonomist.'"

Dr. James E. Lloyd, of the University of Florida, has recently completed a study of flash communication in the genus Photinus, the common smaller fireflies of the eastern United States. (Did you know that the Pacific coastal states, despite their many attractions, have almost no fireflies?) Lloyd, too, found several "hidden" species, first recognized by consistent differences in flash signals, and later found to differ in minor details of body color. In many places two or more species of Photinus fly together, but they are prevented from interbreeding by their different light signals. The males fly at different heights and in different flight patterns; their flashes differ in length, in the number of pulses per flash, and sometimes in the color or intensity of the light. The male is saying, in Lloyd's words: "Here I am in time and space, a sexually mature male of species X that is ready to mate. Over." The female of "species X"

The flash patterns of males of six species of the genus Photinus. Each is signaling to the female of his own species in his own characteristic pattern. No. 1 flies two to four yards high and produces three slow flashes in series. No. 2 flies in a rather straight path somewhat lower, producing single flashes that increase in intensity during emission. No. 3 is a low flier that emits a long flash while executing a lateral curve, while no. 4 makes a series of hops between which he hovers and produces a quick flash. No. 5 is Photinus pyralis with its characteristic J-shaped signal, as discussed in the text. No. 6 is a low-flying species which produces a long flash while jerking rapidly from side to side. (Adapted from a drawing by Daniel Otte in James Lloyd's study of Photinus fireflies.)

responds with a flash at the interval characteristic of her species—as described above for Photinus pyralis. Lloyd was interested in learning how much latitude was permissible without causing a "misunderstanding." In his experiments he used electronic devices for producing artificial flashes of known duration as well as for accurately measuring the response delay of the females. As in the case of crickets—and in fact all "cold-blooded" animals—things happen faster at higher temperatures, so in all his work temperature had to be taken into account.

In any given locality, the males and females are highly attuned to one another's messages; that is, the variation in responsiveness is such that they almost never answer another species. Females occasionally reply once to a flash of inappropriate length, but they do not continue to do so. On the other hand, if one compares the flash signals of species that do not occur together he often finds them to be very similar: here there is no possibility of mistakes being made, and refined "isolating mechanisms" have not evolved. It goes without saying that the integrity of species must be maintained, for interspecies hybrids are generally sterile (like the mule) or at least less well adapted for a specific role in nature.

One would assume that the larger fireflies of the genus Photuris (studied by Barber) always "speak a different language" from the small fireflies of the genus Photinus (studied by Lloyd). This is, of course, generally so, but with some fascinating exceptions. H. S. Barber commented on this as follows:

"Sometimes the familiar flashes of a small species of Photinus male are observed excitedly courting a female, supposedly of the same species, whose flashes appear normal to its kind, but when the electric light is thrown upon them one is startled to find the intended bride of the Photinus is a large and very alert female Photuris facing him with great interest. Does she lure him to serve as her repast? Very often a dim steady light near the ground proves under the flashlamp to be a small, recently killed Photinus being devoured by a nonluminous female Photuris. . . ."

James Lloyd, while working on Photinus, found it possible to obtain females of a given species by walking about in a suitable habitat, imitating the flashes of the males with a flashlight. But now and then the females that signaled back to him turned out not to be Photinus females, but those of the genus Photuris, responding appropriately to specific signals of a certain species of Photinus! Once he watched one of the Photuris females for half an hour and saw her respond to twelve passing Photinus males, in each case after the interval characteristic of that species of Photinus; all of these males were at least partially attracted to her. Finally a male landed near her, and after an exchange of signals ceased to light up after the usual time period. Lloyd checked and found that the Photuris female was clasping the Photinus male and chewing on him. As Lloyd points out:

"The answer to Barber's question has precipitated a deluge of new questions, not the least of which concerns the males of the genus Photuris. Is the female Photuris predaceous before she has mated? If so, how does her mate avoid the fate of attracted Photinus males? . . . Can a single Photuris species prey upon more than one Photinus species with different signal systems? In other words, how many flash patterns do Photuris females have in their 'repertories,' and is predation on Photinus fireflies in any sense obligatory?"

It might be added parenthetically that insects are known that utilize luminescence not for courtship but strictly for luring and then feeding upon various small insects that are naturally attracted to light. Both in North America and in Europe there are certain gnat larvae that spin silken webs close to the ground and emit a dim, bluish light that probably serves to attract tiny midges and other insects into the web. An even better example is provided by the so-called New Zealand glowworm, which is not a true glowworm at all but another gnat larva. These insects live in certain caves in New Zealand and are so spectacular that guided tours are conducted into some of the caves. The gnats lay their eggs in a gluelike substance on the ceiling, and the larvae suspend themselves from silken sheaths and emit a bluish-green light that is said to lure small insects into the tangle of webs, where they are consumed by the larvae. F. W. Edwards describes the experience of entering the depths of one of these caves as follows:

"[After being warned by the guide to be quiet] we stepped cautiously in single file down, down to a still lower level. . . . Then gradually we became aware that a vision was silently breaking on us . . . a radiance became manifest which absorbed the whole faculty of observations—the radiance of such a massed body of glowworms as cannot be found anywhere else in the world, utterly incalculable as to numbers and merging their individual lights in a nirvana of pure sheen."

True fireflies are also capable of remarkable displays at times. Occasionally (especially in the tropics) untold thousands of fireflies will gather in a single tree or several neighboring trees and flash for many hours, sometimes for many nights in succession, producing a glow that can be seen half a mile or more away. Sometimes all fireflies in a tree have been seen to flash in synchrony. Such displays have been reported from Southeast Asia and the East Indies for over two hundred years—but hardly ever from other parts of the world. Hugh M. Smith, while studying the fisheries of Thailand in the 1930's, often took parties of visitors down the Chao Phraya River near Bangkok to observe the displays. In an article in *Science,* he described them in these words:

"Imagine a tree thirty-five to forty feet high thickly covered with small ovate leaves, apparently with a firefly on every leaf and all the fireflies flashing in perfect unison at the rate of about three times in two seconds, the tree being in complete darkness between the flashes. . . . Imagine a tenth of a mile of river front with an unbroken line of [mangrove] trees with fireflies on every leaf flashing in unison. . . . Then, if one's imagination is sufficiently vivid, he may form some conception of this amazing spectacle."

Smith went on to say that the synchronous flashing occurs "hour after hour, night after night, for weeks or even months. . . ." Reports such as Smith's have tended to remove much of the skepticism that greeted earlier accounts. (An author of an article in *Science* some years earlier had attributed the flashing to the twitching of the eyelids, remarking that "the insects had nothing whatever to do with it"!) For years the explanation of this unique phenomenon has intrigued John Buck, of the National Institutes of Health at Bethesda, Maryland, one of our leading authorities on fireflies. Some time ago he found that he could induce synchronous flashing on a small scale in the American firefly Photinus pyralis by using a flashlight at the usual interval of females of this species. When there were many males about, he could sometimes attract fifteen or twenty of them at once, and these would all adjust their flash periodicity in accordance with that of the female. "It is indeed an impressive sight," says Buck, "to see such a group converging through the air toward one point, each member poising, flashing and surging forward in short advances, all in the most perfect synchronism." It seemed possible that small groups such as this might build up within a larger aggregation and so stimulate one another that all fell into synchrony.

In another experiment, Dr. Buck placed a large number of males of this same species in a large, dimly lighted cage, where they soon began to flash in their usual manner. He then subjected the fireflies to sudden and complete darkness, whereupon all of them flashed at once, then again after four or five seconds. The synchrony persisted for some time and then disappeared. Buck felt that the unnatural advent of sudden, total darkness was not of importance in itself, but only because it served to increase the relative intensity of the flashes of neighboring fireflies, causing them to respond to one another's flashes as they would not ordinarily do in nature.

But of course these simple experiments performed on a North American species merely whetted his appetite for the real thing, and a couple of years ago John and Elisabeth Buck took off for Thailand and Borneo. They were successful in finding "firefly trees," and they made photographic and photometric analyses that indicate that synchrony of great numbers of individuals is indeed nearly perfect. They found that (contrary to earlier reports) both males and females occur in these trees, although the females do not participate in synchronous flashing. They showed that mating occurs in the trees, suggesting that the brilliant, synchronous flashes serve as a beacon to attract females from the surrounding forest. This may explain why this phenomenon is most prevalent along rivers in the Far East, for in this part of the world the exceedingly dense, tangled swamps would hardly be conducive to individual flash communication similar to that occurring in a New England meadow. But a "firefly tree" along a watercourse would provide an assembly beacon of ready access. Not only would the synchrony of the flashes increase the brightness but the alternation of light and dark would also be eye-catching, like the flashing neon signs that are a recent invention of man (though I would think that man has overdone a good thing, as he so often does).

The Bucks consider synchronous flashing to be a complex of behavior patterns (congregation, selection of certain trees, flashing, synchrony, and so forth) that have evolved together into a spectacular device for enhancing mating under otherwise difficult conditions. Evidently newly emerged males and females are constantly recruited from the surrounding forests, for individuals do not live more than a few days, and there must be a constant turnover in the population. The Bucks showed that males released in a darkroom are attracted to each other's light, and this suggests that wandering individuals might readily join a flashing swarm. It remains to be proved that there is a traffic of freshly emerged males and females into the trees and of mated females away from them. And it remains to be shown how the males maintain almost perfect synchrony from one end of a large aggregation to the other, when in fact laboratory studies suggest that the males react to one another over only short distances and that their reaction time is considerably greater than the variation in synchrony observed in nature. There is evidence that near-perfect synchrony occurs only in very dense aggregations, while in diffuse gatherings the flashing may be random. In some instances more than one species may aggregate in a given tree, resulting in a complex combination of flashes that is still presumably effective in attracting females of the species involved. All these are matters requiring much further study.

But of course scientists are used to partial and provisional answers; it is their stock in trade, and half the fun of science. H. S. Barber was well aware that his field studies of Photuris were only preliminary. And after a lengthy review of laboratory studies, John Buck concluded:

"In spite of the many morphological and physiological data which concern luminescence in the firefly, there seem to be surprisingly few unequivocal major conclusions which can be drawn."

This is "par for the course." Such is the complexity of living systems that tens of thousands of research workers all over the world each year push our knowledge forward by only a minuscule, with now and then a breakthrough that opens up a new area of ignorance. A century from now our great-grandchildren may marvel at how little we knew about fireflies. At least I hope so. In the meantime we may be unashamedly romantic or unflinchingly rigorous in our attitude toward fireflies, as befits our nature, and still know their magic.

Interlude in the Elysian Meadows: Butterflies

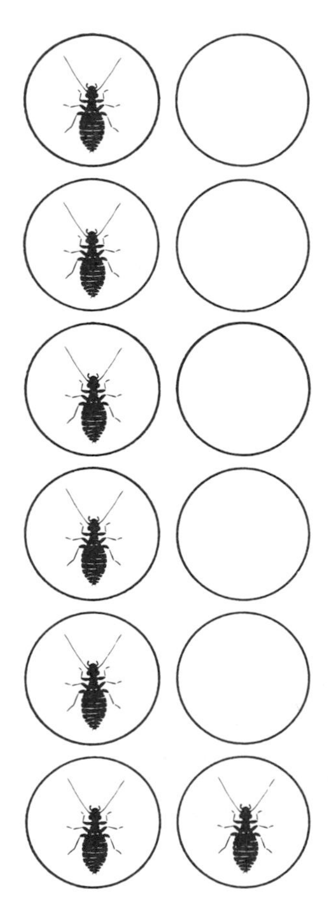

There is a time in every child's life when he has to face up to the fact that he is too old to play hopscotch, to watch Captain Kangaroo, or to eat Sugar Pops: the sweet joys of youth giving way to the more severe world of the adult. Just so there was in human development a time of the wild frontier, of rugged individualism and devil-may-care. Now that man is more sophisticated and there are so many more of him, his world is more regulated and standardized, no matter how well adorned with gadgets. He has learned to do without the virgin prairie and the cry of the eagle, and he will presumably someday learn to do without much of the world that nurtured him. It was his choice to eliminate the wolf, and he would like to eliminate the mosquito and the bark beetle; but in doing so he is losing the untamed forest, the wild lake haunted by a loon. So it must evidently be.

On July 10, 1966, *The New York Times* ran two separate news items on the decline of butterflies. One concerned the small country of Sikkim, on the slopes of the Himalayas, once known as a "butterfly paradise." The Maharajah, it seems, was distressed that the highly successful malaria-control program in that country had resulted in a marked reduction in butterflies. "Today," the *Times* reported, "butterfly hunting, which was once a major occupation and a handsome source of revenue to Lepcha mountaineers, has gone with the DDT." The other article emanated from Washington, D.C., and was headlined "Butterfly Dying as Cities Spread." In both North America and Europe, the bulldozer is charged with eliminating the habitat of many of these insects: "As urban areas spread, pastures and woodlands shrink, and fewer butterflies are seen. . . . Certain species in California are believed to be already extinct; others are slowly declining."

A few months earlier, Dr. Kurt Gohla, Professor of German at Fordham University, spoke to a group at the American Museum of Natural History on "A World Without Butterflies, and One Man's Fight to Delay It." On a recent trip to the Bavarian Alps, Dr. Gohla found that certain areas that once abounded in butterflies are now devoid of them. German biologists suppose that the cause may be found in radioactive dust in the atmosphere or the use of certain chemical fertilizers. Mr. Walther Ender, of Lage, Westphalia, is said to be breeding butterflies in great numbers in the valiant effort to repopulate the landscape.

These items struck home to me, as not long ago I visited some of the meadows where I used to chase butterflies as a child. They were solidly in houses, and the closest approximation to a butterfly was a tattered Kleenex blowing from a trash can.

What good is a butterfly? To the farmer it is an adult cabbage worm or carrot caterpillar, and better off dead. To the entomologist it is a member of a group of diurnal lepidopterans possessing knobbed antennae, a group containing a few pest species but mainly of interest to hobbyists and dabblers. To the romantic poet it is a stray piece of some forgotten rainbow, a vagrant wisp of eternity—but there are no longer any romantic poets to speak of. To the man of the world, the pillar of society, a butterfly is simply nothing at all.

Perhaps, like archy the cockroach, I am really a poet reincarnate. Butterflies matter to me. I remember my first introduction to the tropics: a canyon in Mexico, its floor carpeted with butterflies of many different kinds—a flying carpet, since as I walked along they flew up in a cloud, then settled again behind me. I have since had this experience several times, but I still cannot escape the feeling that it is something rather special, perhaps reserved for that small minority of humans who have not disavowed their kinship with nature. I have luxuriated in the quiet flickering of myraids of Heliconius in the patchy light-and-dark of a tropical forest, and I have chased Parnassius across mountain meadows. I would like someday to meet in person the grandest butterfly of all, the Ornithoptera, or bird-wing, of the East Indies. In his classic *The Malay*

Archipelago, the British naturalist A. R. Wallace wrote of his impressions on encountering his first male Ornithoptera:

"I found it to be as I had expected, a perfectly new and most magnificent species, and one of the most gorgeously coloured butterflies in the world. Fine specimens of the male are more than seven inches across the wings, which are velvety black and fiery orange. . . . The beauty and brilliancy of this insect are indescribable, and none but a naturalist can understand the intense excitement I experienced when I at length captured it. On taking it out of my net and opening the glorious wings, my heart began to beat violently, the blood rushed to my head, and I felt much more like fainting than I have done when in apprehension of immediate death. I had a headache the rest of the day, so great was the excitement produced by what will appear to most people a very inadequate cause."

There is more to be said for butterflies than that they are decorative, and there is more to be seen than the cut and color of their wings. Butterflies were not, after all, designed to please the human eye—they were around long before we were. Their color pattern has evolved as means of communicating with other butterflies and other inhabitants of their world. They are saying: I am a male meadow fritillary, looking for a mate; or I am a monarch, and not good to eat; or I am a leaf, not really a butterfly at all. In fact, the colors of butterflies provide an exceedingly rich field for research into the significance of animal colors generally.

From what we know of the function of color in dragonflies and in birds, we might assume that the adornments of butterflies are primarily involved in courtship and mating and that these are largely "visual animals" that make little use of odor and touch stimuli. But we would be more wrong than right. Unfortunately, the mating behavior of only a few butterflies has been studied in detail. One of the earliest and best-known studies was that made by Niko Tinbergen and his colleagues on a small European butterfly called the grayling. Tinbergen is now a professor at Oxford University, but these studies date from an earlier era when, with a group of enthusiastic students, he studied the behavior of insects on the sands of Hulshorst, Holland. Tinbergen has described these days in a delightful popular book titled *Curious Naturalists.*

The grayling is a small and somewhat inconspicuous butterfly. On the upper surface the wings are brown and ocher, with two round eyelike spots on each wing; on the undersurface they are mottled with gray. Its flight is swift, but upon landing on a tree trunk it quickly comes to rest with its wings closed, blending into the color of the bark. We know from experiments on a number of insects that have color patterns blending into their background that such patterns do, in fact, greatly reduce predation by birds. Evidently the major message the grayling's colors convey is simply: "I am a piece of bark." What of the "eyespots" on the wings? These show up mainly when the butterfly is on the wing, but if a resting grayling is approached closely, it raises its fore wings so that these spots are revealed. The message of these spots might be translated: "Well, you have spotted me; if you must peck, peck here, this is the center of me." Of course, the bird that heeds the message ends up with a small piece of wing, and the butterfly is safe. It is fairly common to find butterflies in nature with beak marks along the fringe of their wings, and there is considerable experimental evidence that small spots near the margin do, in fact, function as "deflection marks," and in many cases enable the butterfly to escape. Hugh B. Cott, in his great compendium *Adaptive Coloration in Animals,* figures a "hair-streak" butterfly with a false head and eyes, and slender "hair-streak" projections resembling antennae, all at the extreme back end of the wings, so that the resting butterfly seems to be facing in the opposite direction than it really is.

On the other hand, the large, showy eyespots on the center of the hind wings of, for example, the io moth, are known to have quite a different function. The io, like most moths, rests with open wings, the front pair lowered to cover the hind pair. When the fore wings are raised suddenly, the large eyespots are revealed, delivering to potential predators the message "I am an owl!" The experiments of David Blest, a student of Tinbergen's, suggest that such eyespots are very effective in frightening off birds. Blest rigged up a small box with a pair of windows in it, with a small light inside. In these windows he could place certain designs that showed up when the light was on. He placed a mealworm between the two windows, then put the box in an aviary. When a bird was about to seize a mealworm, he would turn on the light, exposing the bird to a pair of designs. In general, each of three species of birds tended to be little frightened by bars and crosses, much more often frightened by circles, and especially apt to be frightened off by concentric circles, especially when these were shaded in such a way as to give them a three-dimensional effect.

The grayling, however, relies upon camouflage, a camouflage reinforced by its abrupt manner of landing and remaining motionless on a suitable background. In this respect it does not differ notably from many North American butterflies, including the mourning cloak, the angle-wings, and the satyrs and wood nymphs to which the grayling is related. None of these butterflies is quite so effectively camouflaged, however, as the famous Kallima butterfly of India, which when resting with its wings closed simulates a brown leaf in intimate detail, even to the stem, midrib, veins, and blemishes suggesting diseased spots. The midrib even appears to stand out in relief, though close examination reveals that the wing is flat and that the relief is produced by three closely parallel, contrastingly colored lines, in exactly the same way an artist would produce a three-dimensional effect on a flat canvas. Shading to produce relief, apparent blemishes, and breaks in outline is very common among protectively colored insects, though few have achieved the perfection of the Indian leaf butterfly.

Does this mean that color plays no role in the courtship and mating of these butterflies? This is a question that Tinbergen and his co-workers set out to solve with the grayling. They noticed that male graylings flew up after any butterfly that passed, regardless of the species, sometimes even after grasshoppers, dragonflies, or small birds. However, unless the object proved to be a female grayling they soon returned to their perch. Evidently they first responded to a very general image, but soon were able to identify a female of their own species. Was it by means of the details of her color pattern, which is slightly different from that of the male, and of course quite different from that of other butterflies in the area? Tinbergen tested this by using a series of paper models suspended from a string, and "flown" past a male from the end of a fish pole. He and his co-workers did some 50,000 tests of this nature, extending over several summers. In Tinbergen's words:

"We presented a remarkable spectacle (to say the least) in our practical but scarcely attractive field dress of shorts, broad-brimmed straw hats and sun glasses, each of us with his two rods, all with dangling paper models (some of which were brightly coloured), trying not to get them entangled in bushes or Heather, staring intently after one of our Graylings, trying to follow it on its erratic flight, running after it, suddenly stopping, stalking it, then carefully going through our angling ritual repeated three times, and finally making a few notes; all this with tense, serious faces. No wonder that we drew curious and suspicious glances from the occasional crofter or tourist who happened to see us at work."

It is hardly fair to summarize so much work in a few sentences, but such is the fate of scientific research (when it is summarized at all). Briefly, grayling males were found to respond as well to models made of plain brown wrapping paper as to models painted like the female or even made of the wings of the female. Furthermore, they responded just as well to models made of red, yellow, green, or blue paper, though less well to white and to lighter shades generally. In other words, they behaved as though they were color-blind. Yet the grayling is known to take its nectar primarily from yellow and blue flowers. Tinbergen and his co-workers were able to confirm that this was indeed so. When exactly the same colored papers that had been presented to sexually active males were presented to hungry males, they chose the blue and yellow papers. In short, hungry males responded selectively; sexually active males did not. Tinbergen took this as a further example of a very common phenomenon: in any given situation, an animal tends to respond to only a few of the stimuli it is capable of receiving; these are the "sign stimuli" that release this particular bit of behavior. Evidently the first steps in sexual behavior in the grayling are released by any dark, fluttering object, regardless of its hue or pattern. In later experiments Tinbergen showed that circular or rectangular models were as effective as butterfly-shaped models as long as they moved in the same way. It

Courtship of the grayling butterfly. The "curtsy" of the male (*on the right*) serves to bring the antennae of the female into contact with the scent patches on his front wings. (After Niko Tinbergen.)

appeared that on the whole the male grayling responded best to an object somewhat larger and darker than its own female. The discovery of "supernormal" stimuli such as these in butterflies and other animals has suggested an explanation of the special fascination of Marilyn Monroe and Jayne Mansfield to the human male.

Sexually responsive female graylings typically alight after a short pursuit by the male. The male then begins a complicated courtship. He faces the female, and with curious, jerky movements raises and lowers his closed wings, at the same time holding his antennae out to the side and moving them in a circle. Then suddenly he opens his wings partially and draws them around the tips of the antennae of the female, which are extended forward. The function of this "curtsy" is to bring the scent detectors of the female into contact with scent patches located on his front wings. Once the female has been stimulated in this way, the male moves behind her and mates with her.

Many male butterflies possess patches of modified scales on the wings, and these produce odors that stimulate the female in the final stages of courtship. Some of these odors can be detected by man, and are often very pleasant. According to Tinbergen, the German naturalist Fritz Müller—whom we shall mention again later in this chapter—often carried a live swallowtail butterfly of a certain species about with him on his travels in Brazil, "just for the purpose of sniffing its scent when the mood took him." Professor E. B. Ford, of Oxford University, in his book *Butterflies*, lists the odors of the males of several British butterflies. The common blue is said to smell like chocolate, the clouded yellow like heliotrope, the Bath white like sweet peas, the grayling like "sandalwood, or an old cigar box."

There is a great deal more that might be said about the function of odors in the courtship of butterflies, but suffice it to say that they are much more important—and color pattern less important—than we might have supposed. But the grayling is a relatively dingy species; surely a study of a more vividly colored species would result in a different conclusion. One such study was made a few years ago by Dietrich Magnus, of the Zoological Institute in Darmstadt, Germany, on the silver-washed fritillary of Europe. This butterfly is similar to the great spangled fritillary so common in New England meadows in August: the wings are orange, speckled with dark spots, the undersurface somewhat more dull, washed with silver. Both males and females respond to blue and yellow when seeking nectar (like the grayling) and to green when seeking a place to rest. But the sexually active males pursue various orange or yellow-brown moving objects, even including falling leaves. Only when they are within four or five inches of such an object do they turn away—unless, of course, it proves to be a female silver-washed fritillary, in which case there is an exchange of chemical stimuli followed by courtship flight. Magnus showed the importance of odor by smearing sticks, stones, or his own fingers with the odor of the female. In every case males that were brought close enough would try vigorously to mate with the object, all the while producing their own sexual pheromones.

Dr. Magnus was primarily interested in determining just what factors caused the males to follow a given object; evidently it had to be moving and of a color similar to that of the female, though once again the precise pattern of the female did not appear important. Since he was interested in determining the exact rate of fluttering, shade of color, and size and shape of object most powerful in inducing the male to follow, he decided to prepare a more sophisticated apparatus than the fish pole that Tinbergen and his co-workers had used. He devised the first (and so far as I know the last) butterfly merry-go-round. The merry-go-round had a single boom, about twelve feet long, that could be made to rotate at various speeds. At each end of the boom was an upright rod bearing two pedestals that could also be made to rotate on their own axes at given speeds. When model butterflies were attached to these pedestals, they could thus be made to move forward (actually in a broad circle, of course) and at the same time to flutter. Males readily followed moving dummies of the general form and color of the female, but of course they broke away and flew off as soon as they were close enough to discover that the dummy did not smell like a female. Magnus' remarkable motion pictures of his butterfly merry-go-round in operation proved to be one of the hits of the Tenth International Congress of Entomology in Montreal in 1956. I well remember sitting on the edge of my seat, watching his orange butterflies swirl round and round through clouds of entomological cigar smoke.

Dietrich Magnus' butterfly merry-go-round. On the right end of the boom an artificial female silver-washed fritillary is being pursued by a male attracted from the surroundings. On the other end is a rotating spool with alternating bands of black and orange which has also attracted a male. Dr. Magnus provided no calliope.

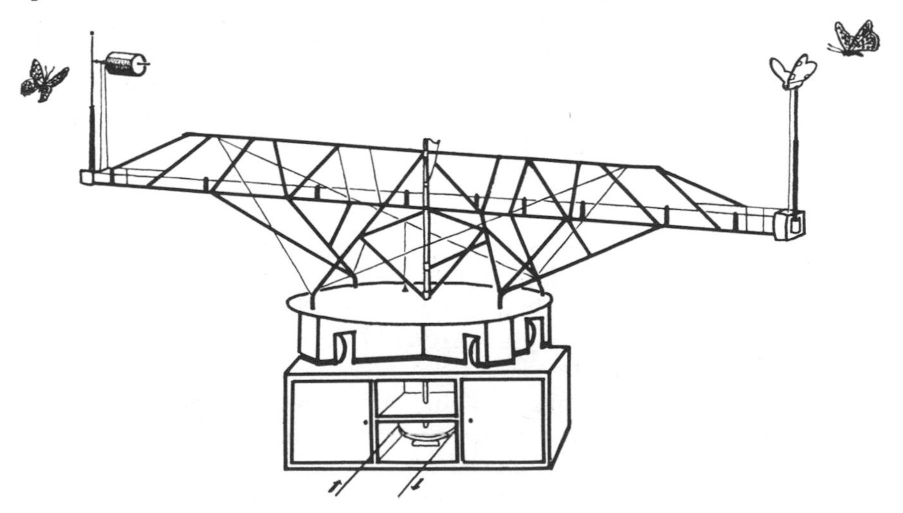

Having perfected his apparatus, Magnus began to manipulate his dummies so as to determine which qualities were attractive to the males. He soon learned that shape made no difference; the males responded as well to circles and triangles as to butterfly shapes. Size was important, however: smaller models released fewer responses than those as large as the female. But models twice the area of the female were even more effective, and models four times the size of the female still better! Also, the rapid alternation between color and noncolor, resulting from wing movements, proved all-important. This information enabled him to turn to a much simpler type of dummy: a mere rotating spool with alternating bands of black and orange. At first he made the orange bands of wings of females, but it was soon apparent that a solid orange color was vastly superior to the patterned wings of the female. When given a choice between a slowly rotating spool and one rotating twice as fast, males always preferred the more rapid alternation of black and orange, up to a rate of seventy-five alternations per second. This is apparently the upper limit of resolving power of the butterfly's eye, and represents a wingbeat much faster than the female is capable of attaining.

Thus Magnus found that an "ideal" female fritillary (in the eyes of the male) is one that is up to four times as large (but of almost any shape), one that is pure orange rather than speckled, and one that flickers its colors as rapidly as the male is able to detect them. Why does not evolution produce such females? Two physical impossibilities present themselves: a butterfly simply does not have the muscles or body form to move its wings that rapidly, and a fourfold increase in the size of the female would entail a similar increase in the male, which in turn would require another fourfold increase in the female, and so on, until there would not be enough vegetation to feed the things. So far as wing shape and pattern go, obviously these have been molded by factors not related to courtship but to the ancestry of fritillaries, to their aerodynamics, and to the advantages inherent in broken patterns and dark wing spots that we have already mentioned. In nature, the male simply does not meet up with these supernormal stimuli. It is just as well that butterflies do not dream. Or do they?

There have been several similar studies showing that pattern is of little importance in courtship and that color is important only in a general way. Miss Jocelyn Crane, of the New York Zoological Society, has spent many years studying butterflies at the William Beebe Tropical Research Station in the Arima Valley of Trinidad. She has become an expert at painting the wings of living butterflies so that they can be released and studied with various abnormal wing patterns. This skill was not easily acquired, for butterflies are delicate creatures, and easily put out of commission by rough handling or the application of foreign chemicals. But once mastered, the skill made it possible to tackle many questions regarding the vivid color patterns of Trinidad butterflies. One of the most striking species, Heliconius erato, proved to be easy to rear in captivity, and Miss Crane sought to find out the significance of the broad band of scarlet that crosses the otherwise black wings of this butterfly. She found that females with the scarlet band blacked out had little success in attracting males. Females with the bands painted green had some slight success, while those with yellow bands had more, those with orange bands still more. In other words, the more the color approached the normal scarlet, the better it worked. Using models, she showed that size and shape of the color badge were of little importance, but that the contrast with the black background generally was. However, her best model was one of solid red.

Heliconius erato. Although only one of many vividly colored tropical butterflies, this black and red species achieved fame as a result of the researches of Jocelyn Crane of the William Beebe Tropical Research Station in Trinidad.

The results with Heliconius erato were especially interesting, since there is good reason to believe that the "red badge of courage" of this species is, in fact, a device for warning predators that this butterfly is not good to eat. Although the majority of butterflies (including fritillaries and the grayling and its kind) are readily eaten by birds, certain kinds, including Heliconius as well as the monarch and its relatives, are known to be distasteful. When Heliconius erato was presented to captured tanagers in a random sequence with butterflies related to the grayling, the tanagers ate all the latter as they were presented, but quickly learned to avoid Heliconius. In fact, half the birds did not even sample the first Heliconius erato presented, indicating that they had already learned in the wild that this bright splash of color meant something that did not taste good. In this butterfly, then, the scarlet badge has two meanings: to birds "I am not good to eat," to butterflies "I am

Heliconius erato." Jocelyn Crane believes that the primary function of the scarlet band is warning and that its role as a sexual releaser has been acquired secondarily.

We know that many brilliant colors—particularly reds and yellows—are associated with unpalatability or the production of venom. This includes not only certain butterflies and many other insects (yellow jackets, for instance) but also vertebrate animals, such as coral snakes, and some of the vividly colored tropical frogs. It was some time, however, before the concept of warning coloration was widely accepted. In an attempt to convince skeptics, Professor G. D. Hale Carpenter, of Oxford University, in 1921 took two monkeys out into the African bush on long tethers and recorded what they found to eat and what they actually ate. In all, they found 244 insects, of which Carpenter considered 143 "conspicuous," 101 "inconspicuous" (one assumes they overlooked many of those that blended especially well with their background). The monkeys ate most of the inconspicuous ones, but less than a quarter of those that Carpenter regarded as conspicuous.

In the 1930's the subject was approached in a slightly different manner by the late Frank Morton Jones, of Wilmington, Delaware, one of the finest American naturalists of his generation. Dr. Jones had the good sense to retire from business at the age of forty-five in order to devote his energies to the study of insects, and the good fortune to live another forty-eight years! An amateur in the very best sense of that word, he became an authority on insectivorous plants, on certain moths and butterflies, and on adaptive coloration. At his summer home on Martha's Vineyard, he established a bird-feeding station on which he regularly placed a wide variety of freshly killed insects in addition to seeds and suet. He watched and recorded the birds that came, and at intervals tabulated the insects that were taken. In all, he used over 4,000 insects of nearly 200 species, noting which were readily eaten by the birds, which were taken later, and which were left on the trays. Of the 31 species that were not usually touched, no less than 24 were brightly colored, chiefly red, orange, or yellow, while all of the more readily accepted kinds were of dull coloration. Further experiments a few years later, in southern Florida, involving very different birds and insects, gave even more convincing results.

Within the past few years, these and related problems have been studied with new experimental approaches and statistical analyses by Lincoln and Jane Van Zandt Brower, a husband-wife team located at Amherst College, Massachusetts. It was the Browers who performed the experiments on Heliconius erato and tanagers mentioned earlier. On a later occasion they transported several blue jays from Amherst to Trinidad, so as to be able to work with birds that had had no prior experience with these butterflies. For ten consecutive days they presented each blue jay with ten butterflies, each a different species, five of them belonging to groups exhibiting warning coloration, the other five belonging to various other groups. Their results were striking. Nearly all the butterflies in the latter group were eaten by the jays. Three of the supposedly distasteful species were in fact highly distasteful, often being pecked or even killed, but rarely eaten. The remaining two were eaten fairly frequently, but more often rejected. The Browers point out that these, and in fact all known "unpalatable" butterflies, belong to only five of the more than twenty major groups of butterflies and that these five groups (in contrast to the remainder) feed in the larval stage on plants that contain acrid or poisonous juices, plants such as the pipe vines, nightshades, milkweeds, and passionflowers. It is the Browers' belief that these distasteful or poisonous substances are taken up by the larvae and passed through the chrysalis stage to the adults, either unaltered or only slightly altered. These substances presumably were evolved by plants as a protection against leaf-feeding animals, but these insects have apparently not only become adapted to feeding on the plants but even to making use of their poisons for their own protection.

Quite recently the Browers have attempted to prove this intriguing possibility. The caterpillars of the monarch butterfly feed upon milkweed, a plant known to contain digitalis-like heart poisons. Birds are known to avoid monarchs, but when forced to feed upon them they become ill, and retch violently. After many difficulties, the Browers developed a strain of monarchs that grew to maturity (with much mortality) on an abnormal food plant lacking these poisons—cabbage. These monarchs proved entirely palatable to blue jays. Since the internal organs undergo almost complete reorganization during the chrysalis stage, the poison is evidently sequestered in the body fluids until the adult stage. It seems unlikely that it is broken down by the larva and resynthesized by the adult, but direct proof awaits the attempt to label the plant poisons with radioactive tracers and follow them through the larva and chrysalis.

Jane Brower's doctoral thesis at Yale University, later published in the journal *Evolution,* was a study of the reputed mimicry of the monarch butterfly by the viceroy, a member of a very different group of butterflies that nevertheless resembles the monarch closely. There has been some argument over this matter—the viceroy tends to be smaller than the monarch, and is often just as common or locally even more common. A good mimic must be less abundant than its model; otherwise birds and other predators will eat the palatable mimic more often than the distasteful model, and fail to learn that this particular coloration indicates distastefulness.

Mrs. Brower's experiments, performed at the Archbold Biological Station in Florida and employing Florida scrub jays, seem to me to settle this matter once and for all. After preliminary tests with eight caged jays, she randomly selected four as controls, and used the other four for her experiments. To each experimental bird she presented butterflies in pairs, one a monarch and the other a butterfly known to be acceptable to the jays. The latter were invariably eaten, while the monarchs were sometimes pecked, especially at first, but more often not touched. After fifty trials she began to substitute viceroys for monarchs, and they, too, were rarely touched. (One of the four birds, however, attacked most of the monarchs and viceroys, though never eating them. Presumably he was a poor learner, or simply a grouch. Or was he a self-appointed apostle of the doctrine that living beings do not react with one another in the manner of chemical equations?)

In the meantime Mrs. Brower fed viceroys to her four control birds—which had not been fed monarchs first—and found that they usually ate them. Her jays were then used in another series of experiments in which monarchs and viceroys were not used. However, after periods of from fifteen to nineteen days her experimental jays were again presented with monarchs and viceroys, and three of them rejected both on sight. It seems impossible to explain these results except as a confirmation of the theories of warning coloration and of mimicry: that animals distasteful to their natural enemies do sometimes acquire brilliant color patterns, that their enemies do in some cases learn to recognize and avoid these patterns, and that other, palatable animals that have evolved similar color patterns thereby gain some protection.

Mrs. Brower's second series of experiments involved the pipe-vine swallowtail and its supposed mimics, the spicebush swallowtail, the black swallowtail, and the black form of the tiger swallowtail. Her results were generally similar to those obtained with the monarch and viceroy: her experimental jays generally did not accept the model or its mimics, while the control jays, not exposed to pipe-vine swallowtails, accepted the mimics. This is a different and more complex situation than that of the viceroy and monarch, however. For one thing, the pipe-vine swallowtail is merely a black species with metallic blue on its hind wings: not really warningly colored in the usual sense. For another, this butterfly is believed to be mimicked not only by the three species Mrs. Brower tested but also by the red-spotted purple, the female Diana fritillary, and perhaps even the male promethea moth, which is a day flier, and much darker than its female. In combination, these mimics often appear more common than their model, and some of them occur north of the usual range of the pipe-vine swallowtail. The fact that these mimics are protected some times and in some places is apparently sufficient to maintain the mimicry. The pipe-vine swallowtail is said to have a strong, disagreeable smell, and presumably it is so very distasteful that birds learn to recognize its rather modest color pattern.

The experimental study of mimicry is very young, and poses complex problems that challenge the ingenuity of biologists. Nevertheless much is being done, and doubts about the effectiveness of mimicry and of protective coloration generally have probably reached an all-time low. There has been much skepticism down through the years, some of it justified, for it is not uncommon for naturalists to overplay an attractive theory. When Abbott Thayer, for example, claimed that the pink colors of flamingos had evolved so as to render them inconspicuous against the sunset, Theodore Roosevelt responded with a long, club-swinging retort, remarking among other things that Mr. Thayer's "suppositions represent nothing but pure guesswork, and even to call them guesswork is a little over-conservative, for they come nearer to the obscure mental processes which are responsible for dreams." Roosevelt, incidentally, applied some very keen insight to these problems back in 1911, concluding that coloration (of birds and mammals, in this case) has no single cause, but arises from "a varied and complex tissue of causes." "As yet," he remarked, "we do not know enough to be able to explain all, or anything like all, the different kinds of coloration and their probable origins." Were he alive today, Roosevelt might be impressed with our progress, but not sufficiently so to retract this statement.

One criticism that has sometimes been brought against experiments using caged birds is that the situation is highly artificial, and may result in greatly altered feeding behavior on the part of the birds. Trying to study these matters under natural conditions is, however, virtually impossible; one simply cannot follow a bird or a butterfly around for long periods of time and record its behavior. Nevertheless the Browers have recently performed a most ingenious series of experiments designed to test the value of certain color patterns under natural conditions. As already mentioned, the male promethea moth is a day flier, dark in color, palatable to birds, and a possible mimic of the pipe-vine swallowtail. Like other giant silk moths (but unlike butterflies), the male finds his mate by means of a potent pheromone discharged by the female into the air, and capable of attracting males over long distances. The Browers purchased a large number of cocoons of this moth and carried them to Trinidad, far from their normal range. When the moths emerged, the females were separated from the males and placed in several trap boxes, so that their scent would be carried off by the wind and so that males that were attracted could enter the box. Males were then released in the morning at an appropriate site, so that many of them could be recollected in the evening in the traps containing the females. Before the

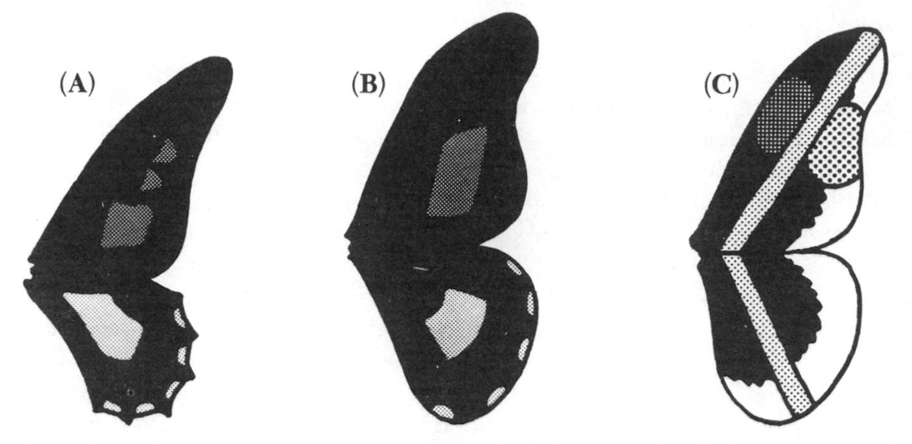

Lincoln and Jane Brower's experiment in "artificial mimicry." *A* is the normal wing pattern of a Trinidad butterfly known to be distasteful to birds; the fore wing has green patches, the hind wing a pink patch in the middle and pink spots along the margin. *B* is a male promethea moth painted so as to resemble this model. *C* is a male promethea with a vivid pattern of red and blue spots and a long yellow bar painted over the normal black and cream pattern—not mimicking anything and in fact made to advertise its palatability.

release, the males were divided into two groups. The experimental males were painted so as to resemble a distasteful, warningly colored butterfly common in the area (such as Heliconius erato), while the controls were subjected to an equal amount of handling and painted with an equal amount of paint, which, however, was black and did not change their general coloration. Thus, if there was a real advantage in "mimicking" Heliconius erato or another distasteful species, more of the artificial mimics than of the black males should be collected in the female cages. That is, the local birds should prey less extensively on moths bearing a badge of color they had learned to avoid.

In fact, in two experiments, in each of which a different Trinidad butterfly was "mimicked," approximately equal numbers of experimental and control moths were collected. The Browers felt that the negative results of these experiments might be a consequence of the fact that their male prometheas were too inconspicuous to attract enough predation so that a difference in the two lots could be detected. They therefore prepared a third experiment in which they painted their experimental moths in a truly vivid pattern unlike that of any other butterfly: a broad yellow band crossing both wings and flanked by red and blue spots. In this instance only about half as many vividly colored moths as black moths reached the female cages. Evidently the bright color did indeed attract the attention of birds, but this was a pattern they had not learned to avoid. At least it was established that predation by birds is significant and that color pattern can influence the incidence of predation.

After this limited success, Lincoln Brower returned to Trinidad for another summer, where he and his associates repeated and extended these experiments. After releasing 1,246 male moths over 24 days, they noted that those painted as "artificial mimics" seem to have an initial advantage, but within a few days they were eaten more commonly than the nonmimics, as before. But when about 400 males were released over 3-day periods only, in 6 separate series, some 26 per cent of the mimics arrived safely at the cages containing females, only 19 per cent of the nonmimics. Evidently the mimetic pattern was indeed advantageous; but Brower and his associates had previously been flooding the area with so many artificial mimics (in relation to the number of natural models) that the birds learned, over a period of days, that this particular pattern usually went with a thoroughly palatable meal.

The Browers, Frank Morton Jones, and others have studied birds as predators for the obvious reason that (despite some arguments to the contrary) they do prey extensively on butterflies and are influenced by warning colors and mimetic patterns. Some critics of mimicry have pointed out that there is little evidence that predatory insects and spiders are influenced by such coloration—and they are perfectly right. Preying mantids accept monarchs and viceroys readily, and some assassin flies prey upon wasps as a regular diet. In my work on solitary wasps I have often observed that these insects are not at all deceived by mimics. Philanthus bicinctus, a large and handsome wasp I have watched preying upon bumblebees in great numbers in Yellowstone National Park, is simply never "fooled" by several species of flies that occur in that area and resemble bumblebees closely. Conversely, Bembix pruinosa, a fly predator I discussed at some length in *Wasp Farm*, very often preys upon the drone fly, which is such an effective mimic of the honeybee that the ancients failed to distinguish them, and believed that honeybees arose from filth—the breeding place of drone flies. For toads, the Browers showed the drone fly to be a highly effective mimic of the honeybee. Most workers would agree that protective coloration of all kinds is mainly effective against vertebrate predators, most of which hunt with their eyes and have large enough brains to profit by their experiences.

Thus far we have been discussing cases of mimicry in which one or more palatable species have evolved color patterns that resemble those of a distasteful or venomous species. This type of mimicry has been recognized for a long time. Kirby and Spence, in their *Introduction to*

Entomology, published in 1817, mentioned the resemblance of certain flies to bees. But it was Henry Walter Bates who first developed the theory fully, and for this reason resemblance of this type is termed "Batesian mimicry." As a young man in Leicester, England, Bates had been apprenticed to a hosiery manufacturer and had later been a clerk in a brewery. He was soon corrupted from following such useful pursuits, however, by a natural love of insects and by reading such books as Charles Darwin's *Voyage of the Beagle* and W. H. Edwards' *A Voyage up the River Amazons.* He also made friends with a local English master, Alfred Russel Wallace, and the two were soon plotting their own voyage of exploration. In 1848 they set sail for Brazil, where Wallace was to remain four years, Bates eleven. When Bates returned to England, Darwin and Wallace had already presented to the Linnaean Society their theory of evolution by natural selection, and Darwin's revolutionary book *On the Origin of Species* was shortly to appear.

Bates' extended studies in Brazil, later set forth in fascinating detail in his classic *The Naturalist on the Amazons,* included many observations on various insects that resembled their background and others that resembled specific objects such as twigs and the droppings of birds. He also recognized that certain butterflies bore a close superficial resemblance to unrelated, more common species, especially those related to Heliconius, which he correctly assumed to be distasteful. He considered such mimicry as simply another form of adaptive coloration, serving to decrease attacks by predators. Darwin was quick to recognize the importance of Bates' work, and wrote to Bates soon after his first paper on the subject appeared in 1862:

"I have just finished, after several reads, your paper. In my opinion it is one of the most remarkable and admirable papers I ever read in my life. The mimetic cases are truly marvellous, and you connect excellently a host of analogous facts. . . . I rejoice that I passed over the whole subject in the *Origin,* for I should have made a precious mess of it."

A few years later Wallace contributed further examples from his work in the East Indies. Then, in 1878, a versatile German scientist, Fritz Müller, expanded upon Bates' work in an important way. Müller had been trained in medicine and physiology, but he was fascinated by evolution, and, like Bates, spent many years traveling in that wonderful natural laboratory, Brazil. Bates had noted that in some cases his mimics were distasteful, like the models, and he was puzzled by this. Müller proposed that since predators must learn to recognize unpalatable species by their color patterns, it is advantageous for several distasteful species to assume a common color pattern, since in this way the predators have fewer patterns to learn and the exploratory predation is spread over several species.

We now know that "Müllerian mimicry," as it is called, is by no means rare. Heliconius erato, for example, has a "twin" species on Trinidad; the two are equally distasteful to birds, and difficult to tell apart, although in fact not closely related. The same is true of several other Trinidad butterflies and in fact many tropical butterflies, beetles, and wasps. In some instances there are several Müllerian mimics, all distasteful in varying degrees, and these have several palatable Batesian mimics, so that a common color pattern carries through many different, unrelated species. One of the most striking examples in the American tropics is the so-called "transparency group": a diverse assortment of butterflies and day-flying moths all of which have transparent wings except for limited areas of dark coloration. This phenomenon was first described by Bates, who knew of seven species, but now we know of nearly thirty species that enter this complex of Müllerian and Batesian mimics. The curious thing is that the transparency has been achieved in different ways in different groups: in some the scales on the wing have been reduced to fine hairs that permit the light to pass between them; in others the scales remain large and dark but are greatly reduced in numbers; in others they are greatly reduced in size. The moths entering this complex employ still other means of achieving transparency: in one group the scales are tilted on their edges, so the light passes between, while in another the scales have themselves become transparent. It has never been my good fortune to visit an area where these transparent butterflies and diurnal moths occur. They must provide a striking vision of otherworldliness as they flicker through the rain forest—a sight no bird is likely to forget, once having associated it with an unappetizing flavor.

Some of the largest mimetic complexes center upon certain of the tropical social wasps. Several species of virulent stingers that live in large colonies may share a similar color pattern, and this pattern is also assumed by various solitary wasps and even by certain flies and day-flying moths, the latter even having acquired a "wasp waist" as well as clear wings and a "waspish demeanor." The "tarantula hawks," large solitary wasps of the American tropics and southwestern United States, also form the center of a complex of orange-winged Müllerian and Batesian mimics that includes other wasps, several kinds of flies, and even a beetle and a grasshopper. To make a grasshopper look like a wasp takes some doing, but in fact it is such a striking mimic that many a wasp collector has been fooled. And I remember by shock when, after I had pursued a tarantula hawk across a Mexican pasture, I discovered in my net nothing but a large blue beetle that had folded its orange wings beneath its wing covers.

In more northerly climates we are less aware of such things. However, the value of a common warning color pattern may explain why so many different species of yellow jackets adhere to the same black-and-yellow pattern; and this pattern is emulated by a number of similarly colored but completely harmless flies. Incidentally, it has been shown that some of these mimetic flies have almost exactly the same wingbeat frequency as their models; thus they also sound like yellow jackets. It is, of course, a requisite of mimics that they occur in much the same habitat and behave very much like their models. A perfect mimic would look, smell, sound, and perform exactly like its model—but unless it did something different mimic and model would likely become pretty confused themselves!

This raises the question: How much does a mimic need to resemble its model in order to gain protection? One of the objections sometimes raised against the theory of mimicry is that in its early stages (when a viceroy was just beginning to evolve toward the color of the monarch, for example) it would confer no advantage, and mimetic patterns could therefore not accrue gradually by natural selection. But Jane Van Zandt Brower found that her scrub jays, which had learned to reject monarchs and viceroys, also rejected the much darker Florida race of the viceroy, which mimics another distasteful butterfly, the queen; they even rejected the queen, which to a human observer looks quite different from the monarch. She remarks that this tendency of birds "to generalize from experience" suggests that "even a broad similarity may be of protective advantage and that an incipient mimic would thus be favored over its non-mimetic variants." Students of mimicry now agree that even very slight resemblances may confer slightly higher survival rates, at least during periods of prey abundance. We have to remember that a starving creature will eat almost anything; Hugh Cott points out that during times of war the Germans have been known to make coffee from acorns and Parisians have been known to eat rats. However, such situations are, fortunately, very rare.

We now know that there are cases of imperfect mimicry that are maintained in nature for very good reasons. George Stride, while at the University of the Gold Coast in Africa, noted that a local butterfly, Hypolimnas misippus, was actually an imperfect mimic of an orange-brown relative of our queen butterfly, sometimes called the African queen: the latter has on each of its hind wings a large white patch that is lacking in Hypolimnas. Dr. Stride decided to construct a "more perfect" mimic by making a scrambled butterfly, using the body and fore wings of Hypolimnas misippus and the white-blotched hind wings of the African queen. However, the male Hypolimnas declined to court it. But when he made a second scrambled butterfly consisting of the body and fore wings of the African queen and the hind wings of Hypolimnas, the males of the latter actively courted it. Still further experiments showed beyond doubt that the white patches on the hind wings serve as a sexual inhibitor to Hypolimnas males. White is known to be unattractive to males of many butterflies, for reasons we do not understand, and it appears that "white inhibition" is so ingrained in males of Hypolimnas misippus that it has prevented this species from achieving a more perfect mimicry. Obviously, the males of the model are not bothered by white inhibition.

Incidentally, the males of Hypolimnas misippus are not mimics at all, but have a very different pattern of black and white. In fact, males of many Batesian (but not Müllerian) mimics retain the pattern of their own group rather than adopting the mimetic pattern of the female. Thus while both sexes of the North American red-spotted purple have evolved color patterns that resemble those of the pipe-vine swallowtail, only the female Diana fritillary has done so; the male is radically different from the female, and has an orange-and-brown pattern not unlike that of other fritillaries. In the case of the tiger swallowtail, the female occurs in two color forms, one yellow and black-striped like the male, the other mostly black, and an effective mimic of the pipe-vine swallowtail. The black form is common in areas where the pipe-vine swallowtail is common, but in New England, for example, where the model is rare, only the yellow and black "tiger" pattern occurs commonly. Yet the "tiger" pattern occurs in some females in all parts of the range, and one wonders why, if the dark female is at an advantage because it is an effective mimic of a distasteful species, the yellow and black-striped form is retained in the southern parts of the range. Why not the solution of the Diana, in which all females are mimetic; or that of the red-spotted purple, in which both sexes are mimetic?

The usual answer is that while the black female tiger swallowtails do have greater survival rates, the "tiger" patterned females have greater mating success—that is, the males find a pattern more like themselves and more like the original color pattern to be more attractive. Thus there are two opposing pressures on the female population: greater reproductive success balanced against greater survival capacity, and this is sufficient to maintain both forms in a "balanced" state. Until recently there was little real evidence to support this conclusion. In 1965 John M. Burns and a group of his students at Wesleyan University, Middletown, Connecticut, undertook to collect large numbers of tiger swallowtails in Maryland and Virginia, and to dissect these in order to find out how often they had mated. This is not so difficult as it sounds (though collecting several dozen female tiger swallowtails is); the sperms are transferred to the female in a packet, and the walls of the packet persist in the female. The tiger swallowtail usually mates several times. Dr. Burns found that the 28 tiger-patterned females he studied averaged about two sperm packets per female, but 85 dark, mimetic females averaged about 1.6 per female. In other words, it appeared that in these areas the dark females were only about 80 per cent as successful as the tiger-patterned females in attracting mates. Apparently the tiger swallowtail has been subjected to the same conflicting selection pressures as Hypolimnas, studied by Stride in Africa, but it has responded in a different way.

Probably it is usually the importance of courtship signals that prevents males from becoming mimetic. It is known that males of certain butterflies, when rendered colorless, are less successful in mating. We believe that in many species there is an actual exchange of visual signals between male and female, so a color deviation on the part of the male may result in less success in mating—the female having the power of refusing him. This is a very critical matter on the side of the male, who after all serves no function other than inseminating the female (of course, I am talking about butterflies, and don't mean that as a universal rule). The female, on the other hand, must survive well beyond mating, and lay her eggs. We still know very little about the importance of male color pattern in stimulating female acceptance; this is one of many aspects of this subject waiting to be explored. Lincoln Brower concluded a recent review of this subject with the statement:

"It seems most probable that the strength of female visual selection is great enough in Batesian groups to prevent mimicry in the male, but in Müllerian groups the increased importance of scent has lowered the visual disadvantage sufficiently so that it is outweighed by the benefits accrued from mimicry."

It should perhaps be added that Batesian mimics other than butterflies are normally mimetic in both sexes: for example, flies that mimic bees and wasps, spiders and beetles that mimic ants, and so forth. In the final analysis there are probably more cases of mimicry among flies, wasps, beetles, and spiders than among butterflies, but butterflies will always remain the major subject of studies of coloration, and justly so, for they are larger and are better experimental animals than many other insects. And somewhere in the world there is a butterfly that exhibits almost any conceivable situation one could dream up, and some that he couldn't.

Consider, for example, the African swallowtail Papilio dardanus, a species dear to the heart of E. B. Ford and his co-workers at Oxford University. This large butterfly occurs over all of Africa south of the Sahara, as well as on Madagascar. The male has the "swallow tails" well developed and is mostly yellow, with a black pattern along the outer wing margins; however, the undersurfaces of the hind wings, which are exposed when the butterfly is resting, are colored to blend with the background. The males show only slight variation throughout their broad range; but the females are another matter. On Madagascar they are very similar to the males, but on the continent they lack tails (except in Ethiopia, at the northern edge of the range), and fall into five different geographic races. Furthermore, within each race the females assume up to three quite different "disguises," each mimicking a different model. One color form has a black-and-white pattern, mimicking a distasteful species with a similar pattern, another a black-and-yellow pattern like that of another unpalatable species, still another an orange-brown pattern mimicking the very same African queen that is mimicked by the very different butterfly that George Stride studied in the Gold Coast. Presumably, by mimicking three different models, each mimic is able to remain less abundant than its model, while the total population of the species may be fairly high. To further complicate the picture, nonmimetic females occasionally show up, and in Kenya, where the models are rare, various intermediates and imperfect mimics are common.

This very complex array of color forms, all within a single interbreeding species, posed a challenge and opportunity to biologists interested in learning more about the inheritance of mimetic color patterns. C. A. Clarke and P. M. Sheppard, of the University of Liverpool, arranged to have live butterflies shipped to them by air from various parts of Africa. They were able to breed them in greenhouses in Britain and to cross the various races and color forms at will. They found that these strikingly different forms had very simple hereditary mechanisms. Apparently each new mimetic form arises first as a small difference in color pattern resulting from a simple genetic change, that is, a mutation at one point on a chromosome. If this change results in a slightly higher survival rate as a result of incipient mimicry, other genes that modify its effect so as to make the butterfly a better mimic will be favored. The population will then consist of the original nonmimetic form and the new mimetic one. If another mutation occurs at the same locus in the original form, making it look something like another model, this too will be favored, and will accumulate modifiers that enhance its effect. Thus each mimetic form, as well as the original nonmimetic form, will be determined by a single gene and its modifiers, this gene capable of "switching" from one form to another. When Clarke and Sheppard crossed stock of two mimetic forms occurring in the same area, the result was one or the other mimic; but when they crossed a mimic with stock from a different part of the range, the offspring were intermediates of various kinds, since the genetic background—that is, the modifiers—were very different. Their work confirmed a theory proposed many years earlier: that mimicry, even though it may involve spectacular differences in color and even in behavior from nonmimetic stock, nevertheless arises from a simple genetic change that is gradually reinforced by other changes enhancing its effect.

With the development of better methods of rearing, shipping, and storing butterflies, of studying them in large cages, of marking them in the field, painting their wings, and so forth, there seems hardly any limit to what we may hope to learn in the future. Not long ago I visited Lincoln Brower's laboratories at Amherst College. He has a greenhouse full of native and exotic species of milkweeds as experimental foods for his monarch butterflies, as well as large rearing cages for butterflies with carefully controlled temperature, humidity, and light. Caged birds are being presented with butterflies by automated devices, and their reactions recorded on video tape—completely removing the distraction of the human experimenter, to say nothing of the tedium. But perhaps the most remarkable item was Dr. Brower's refrigerator—bulging with untold thousands of deep-frozen butterflies ready to be used in his experiments.

As I say, there is hardly any limit to what we may hope to learn in the future—and hardly any chance of striking the "bottom of the barrel" of knowledge in this field. Provided, of course, we still have plenty of butterflies and at least a few energetic people studying them! The fact is that every species is a study in itself; as Aristotle taught a very long while ago: "We must take species separately, and study the nature of each." The coloration of every species is a compromise between sundry factors that have influenced its economy, past and present. Natural selection favors now concealment, now mimicry, now more effective courtship signals. The different wings and wing surfaces, or parts of them, may tell different stories. Males and females may look quite different, and the females may even exist in two or more quite different color forms. Henry Walter Bates sensed all this more than a century ago, when he remarked in *The Naturalist on the Amazons*, regarding the wings of butterflies:

"On these expanded membranes nature writes, as on a tablet, the story of the modification of species, so truly do all changes of the organisation register themselves thereon. . . . As the laws of nature must be the same for all beings, the conclusions furnished by this group of insects must be applicable to the whole organic world; therefore the study of butterflies—creatures selected as the types of airiness and frivolity—instead of being despised, will some day be valued as one of the most important branches of Biological science."

This new knowledge of butterflies—and of the depths of our ignorance about them—does nothing to decrease their attractiveness. Indeed, the tiger swallowtails that (all too rarely) drift past my back porch are the more delectable for my knowing that a few hundred miles south, many of them are "black tigers"; and for my knowing something about the origin and function of mimicry. Let us hope that the study of butterflies will never be left entirely to the professors. I judge from reading Urquhart's book *The Monarch Butterfly* that there are still many enthusiastic butterfly lovers and that some of them have been most helpful in recent studies of the migration of the monarch (which is quite another story, not followed here). E. B. Ford, in his book *Butterflies*, suggests several ways in which persons with a little knowledge of genetics and population dynamics can share the fun of discovering new knowledge. In the study of butterflies, both amateurs and professionals may be reminded that, in William Morton Wheeler's words, the world of nature is "an inexhaustible source of spiritual and esthetic delight." And Professor Wheeler goes on, in the concluding paragraph of one of his best-known essays:

"Our intellects will never be equal to exhausing biological reality. Why animals and plants are as they are we shall never know, of how they have come to be what they are our knowledge will always be extremely fragmentary, because we are dealing only with the recent phases of an immense and complicated history, most of the records of which are lost beyond all chance of recovery; but that organisms are as they are, that apart from the members of our own species they are our only companions in an infinite and unsympathetic waste of electrons, planets, nebulae, and suns, is a perennial joy and consolation. We should all be happier if we were less completely obsessed by problems and somewhat more accessible to the esthetic and emotional appeal of our materials, and it is doubtful whether, in the end, the growth of biological science would be appreciably retarded. It quite saddens me to think that when I cross the Styx I may find myself among so many professional biologists, condemned to keep on trying to solve problems, and that Pluto, or whoever is in charge down there now, may condemn me to sit forever trying to identify specimens from my own specific and generic diagnoses, while the amateur entomologists, who have not been damned professors, are permitted to roam at will among the fragrant asphodels of the Elysian meadows, netting gorgeous, ghostly butterflies until the end of time."

Paean to a Volant Voluptuary: The Fly

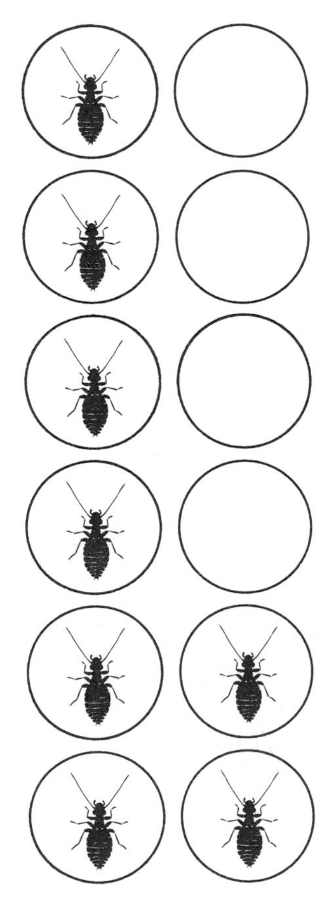

The fly has so long been an intimate of man that we take him pretty much for granted, like taxes or mothers-in-law. One of our best-known folk songs is about "Ole Massa" who was killed when his horse was "bitten by de blue tail fly," while another concerns the death of "The Old Lady Who Swallowed a Fly." The latter might still be alive if she had heeded the Spanish proverb "En una boca cerrada no entran moscas" (Flies don't enter a closed mouth); but then, of course, we would not have such a gruesomely delightful song to sing. Gruesomely delightful: that is rather the story of flies; but on the whole we don't know them well enough to share their delights.

Man has had flies as his unwanted companions every since the dawn of civilization, and doubtless before. In the words of the British entomologist Harry Eltringham:

> Profound research does not disclose
> The epoch when the fly arose
> To plague creation.

The Philistines made obeisance to Beelzebub, the god of flies, in the hope of propitiating him and so reducing the curse of flies. Malcolm Burr, in his book *The Insect Legion: The Significance of the Insignificant,* speaks of this as "the first historical account of the appointment of a Fly Control Officer." Then came the ten plagues of Egypt, six of which (according to Burr) were entomological, and at least two and perhaps four of which were attributable to flies. Fly-borne diseases are said to have hastened the decline of both Athens and Rome. Sleeping sickness long delayed the civilizing of Africa, and yellow fever the opening of the American tropics. And flies are still with us, as a trip to the arctic or the tropics, or to one's own back porch, will confirm. Anyone not fully convinced is invited to visit the Adirondacks in June or a St. Louis alley in September. Studies some years ago in China, by the way, showed that in slum districts the average fly carried 3,683,000 bacteria. But take heart: in relatively clean communities he carried only 1,941,000 bacteria.

Flies are more enjoyable, to be sure, if one doesn't look at them quite that closely; just closely enough to appreciate how elegantly they are designed for reveling in the world's filth and backwaters and for carrying the act of reproduction to unimagined heights. To the sophisticate, to the world-weary, I especially commend the fly, fleeting yet eternal, past master at the fundamentals of living.

I use the word "fly," of course, as an entomologist would use it, and as he feels everyone should use it: to mean a member of that very large group of insects that develop from legless maggots or wigglers and have, as adults, only one pair of wings. Originally the word referred to any flying insect, and in this way it has come down to us in such words as dragonfly, mayfly, firefly, ichneumon fly, and the like—insects that have four wings and are thus not "true" flies in the modern sense. Unfortunately, the meaning of the word has been still further compressed, so that to many people the word "fly" refers only to the housefly and several similar kinds that occur commonly around homes. In fact, gnats, midges, and mosquitoes are flies (the word "mosquito" being the diminutive of the Spanish word for fly, *mosca*), as are many others bearing no close resemblance to the housefly: insects such as crane flies, black flies, robber flies, and a diverse assortment of others. Harold Oldroyd has recently outlined the world of flies in a beautifully illustrated book, *The Natural History of Flies.* Oldroyd, who is senior fly specialist at the British Museum, has probably had more flies cross his desk than anyone else in the world. At first glance this may seem a dubious distinction, but readers of his book are likely to conclude that it is, after all, a rare and exciting privilege. As I wrote, in reviewing Oldroyd's book for *Scientific American:*

"[This] book is particularly recommended for those who so eagerly study the photographs of Mars for signs of life. There may not be much of anything on Mars, but on earth there are some 80,000 species of flies, many of them little known and most of them a good deal stranger than the organisms that fill the dreams of the exobiologists.

"Have you heard, for example, of the downlooker fly, which sits head down on the trunk of a tree 'as if it were looking for a victim'—when in fact no one knows what it feeds on? Or of the coffin fly, which maintains itself through many generations in human bodies buried in coffins, although no one knows how it gets into them? Or the petroleum fly, known only from pools of crude petroleum in California, where it lives on other insects that became trapped in the oil? Or of mosquitoes of the genus Malaya, which subsist by filching honeydew from the jaws of worker ants? There is hardly a page in Oldroyd's book that does not introduce the reader to one fantastic fly or another, and hardly a paragraph that does not contain a confession of ignorance and an implicit plea for more studies on the life histories of these ubiquitous but little-understood animals."

What is the essence of a fly? (I am referring to its basic features, of course, and not to the aroma emanating from a fly-blown carcass. The latter also tells us a good deal about the character and success of flies, but makes it hard for us to approach them dispassionately, or for that matter to approach them at all.) Since it is their two-wingedness that distinguishes flies from other insects, we should consider the significance of this. Most of the more advanced kinds of insects have developed a mechanism for reducing the two pairs of wings to a single functional unit: in bees and moths the front and hind wings are hooked together, and beetles fly only with the hind wings, the front wings being used as protective covers. The flies have carried this trend to the ultimate. The middle segment of the thorax, which bears the first and only pair of functional wings, is greatly developed at the expense of the other two segments, forming a robust box crammed with powerful flight muscles. These muscles are of the indirect type; that is, they move the wings by changing the shape of the thorax itself. They are also of the asynchronous type, as discussed in Chapter 4. This means that (in contrast to dragonflies and many other insects) the rate of wingbeat is independent of, and very much faster than, the nervous input. David S. Smith, of the University of Virginia, speaks of muscles of this type as "the most spectacularly active tissue that animals have evolved."

Apparently the rate of contraction of the flight muscles is determined by rhythms intrinsic to the muscles but influenced by the area of the wing. That this is so can be shown by trimming down the size of a fly's wings: he will then beat them more rapidly. You will, of course, need a stroboscope to determine this. This instrument can be set to produce a flash of light a certain number of times a second. When the rate of flashing is the same as the wingbeat, the wings appear stationary. Stroboscopic study has revealed that the normal rate of wingbeat of flies exceeds that of all other insects. The small fruit fly Drosophila, so dear to geneticists, beats its wings 250 times per second (considerably faster than the honeybee), while mosquitoes beat their wings up to 600 times per second, certain small midges over 1,000 times per second. The speed of flight does not necessarily correlate closely with wingbeat (midges and mosquitoes have very narrow wings, and the rapid wingbeat to some extent compensates for the small wing area). Nevertheless it is probably true that the fastest insects are flies. Many years ago C. H. T. Townsend claimed that the deer botfly is able to fly 815 miles an hour. Townsend was such an enthusiast for flies that he wrote a "manual" on flies that ran to twelve volumes! It must be admitted that no one has confirmed his observations on the speed of the botfly or come anywhere near it. Some of the larger horseflies, unquestionably among the strongest fliers, are now believed to average only about thirty miles an hour. Such flies have, however, been seen to circle an automobile going forty miles an hour and then alight on it, so apparently they are able to fly at considerably higher speed in short spurts. Many flies are also hoverers par excellence; that is, they are able to remain stationary in the air for long periods while beating their wings rapidly and often producing a characteristic hum or whine. Many are able to fly backward with ease or to perform acrobatics that would shame the Blue Angels: witness a parasitic fly laying its eggs in the abdomen of a bee while the latter is in flight or a housefly making a somersault to land on the ceiling.

I said that flies have only a single pair of wings, but this is in the nature of a half-truth, and setting the record straight may help to explain some of their aerial derring-do. In fact the hind wings are present, but in the form of a pair of short stalks each terminating in a knob. Early observers supposed the fly used these a bit as a tightrope walker would use a pole weighted at each end, so they called them "balancers." We now know that this is not their function, so we call them "halteres," which connotes essentially nothing, and allows us to change our minds from time to time regarding their function. The halteres move rapidly during flight, and have, near the point of articulation with the body, an extensive series of small sense organs. It has been known for a long time that if one removes the halteres from a housefly and then releases it, the insect goes into a spin and eventually crashes on the floor. However, if one attaches a short thread to the tip of the abdomen of a haltereless fly, its flight is greatly improved. Apparently the thread acts as a stabilizer, a bit like the tail of a kite. Evidently the halteres normally act as stabilizers, but they cannot act as "balancers," since their weight is far too small to be effective in this way, and especially since a fly with only one haltere removed—and therefore way "off balance" if weight were important—is able to fly almost normally.

Extensive studies by J. W. S. Pringle, of Oxford University, have shown that the sense organs at the base of the halteres are all-important. These are stimulated by distortions of the cuticle during the oscillations of the halteres, and especially by abnormal stresses during turning movements. Thus the halteres "inform" the fly of any incipient instability, and permit it to correct its flight accordingly. In the words of Talbot Waterman, of Yale University, the halteres "may be considered as analagous to the turn indicator of an airplane in which the precession of a

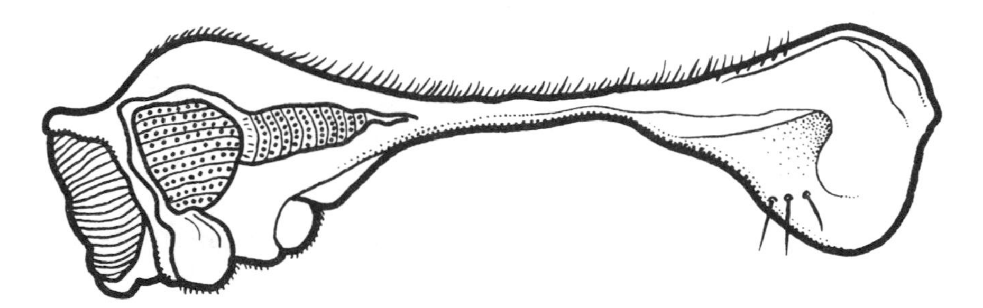

The haltere (modified hind wing) of a fly. The point of attachment to the body is at the extreme left, next to it a complex sense organ that detects distortions in the surface as the haltere moves, informing the fly of any irregularities in flight.

gyroscope is used to signal the direction and rate of turn." Waterman characterizes this as the only known use of the properties of a gyroscope in the animal world. Wind-tunnel experiments have also shown that the last segment of the antennae of the housefly provides an effective airspeed indicator operating on the same basic principles as those of aircraft. In a wind tunnel, as in free flight, a suspended fly contracts its leg muscles so as to "retract its landing gear," again paralleling a very much later invention of man. If only our aircraft were as agile and dependable as flies; and if only a pair of them could produce several thousand more of their kind in a few weeks, as a housefly can!

It is the uses to which flies put their wings that especially intrigue us (and here the fly shows himself a good deal saner than his very recent imitator). There is much more to the story than merely flying around and looking for a mate or a source of food. For example, flies produce sounds with their wings, sometimes only weakly audible to us, but often highly important to the fly as a means of communicating simple messages. The tone is a function of the rate of wingbeat, and responsiveness to a particular tone may help flies to aggregate or to find the opposite sex of their own species. The fact that male mosquitoes are attracted to the sound of the wings of the female has been known for a long time. A few years ago there was much publicity over the possibility of trapping mosquitoes by playing them recordings of their own "love songs" ("Hot Platters by Mosquitoes," one article was titled). Unfortunately, the females are not attracted to the sound of their own wings, and the mosquito is one of numerous biting flies in which only the female bites. If one could trap 100 per cent of the males it would be fine, but trapping 95 per cent of them merely awards to the remainder a particularly orgiastic existence, while leaving us with just as many mosquitoes in the next generation.

The role of sound in the reproductive behavior of the mosquito has been investigated especially by Louis Roth, whom we met as a cockroach expert in Chapter 3. As a graduate student at Ohio State University some years ago, Roth had not yet succumbed to the charms of the roach, and undertook a detailed study of the yellow-fever mosquito. Roth found that these mosquitoes start to mate while in flight, but often come to rest, the male beneath the female in a "face-to-face" position; the genital organs remain fastened together for anywhere from four to fifty-nine seconds. Both males and females tend to mate several times. The best way to induce mating in the laboratory, Dr. Roth found, is simply to shake the cage and make them take flight: it is the sound of the female in flight that stimulates the male to copulate.

Curiously, the male yellow-fever mosquito responds to a considerable range of sound frequencies, and will therefore attempt to copulate with mosquitoes of other species. Apparently there is no close-range odor stimulus to help the male recognize his own species, but for mechanical reasons (differences in body form and structure of the genital organs) he is simply unable to mate successfully with females of other species. The male yellow-fever mosquito responds beautifully to the sound of tuning forks having frequencies of anywhere from 250 to 500 cycles per second. When such tuning forks are struck outside the gauze netting of the cage, the males fly to this side, seize the netting (or sometimes another male), and actually attempt to copulate with it. The stark simplicity of this response is in strong contrast to the interchange of delicate odor and sight stimuli occurring among butterflies.

The responsiveness of males to a broad spectrum of sound frequencies is shown by the fact that some persons have found themselves attracting flies with their voices. Jaakko Syrjämäki, of the University of Helsinki, has the habit of humming a certain Finnish folk song while working. But while conducting research recently on the swarming of midges, he found that the initial *G* of the song suddenly drew the swarm to his mouth! The midges responded well to *F* and *A* also, less well to *E* and *B*. It would seem that persons in the presence of male midges and mosquitoes should be especially alert to the Spanish proverb cited in the opening paragraph of this chapter.

It has been suspected for a long time that the antennae of the males, which are very much more bushy than those of the females, play an important role in detecting sounds. It was in 1855, in fact, that a Baltimore physician, Christopher Johnston, described the "auditory capsule" at the base of the mosquito's antenna, surmising that by means of this structure the male was guided to the "sharp humming noise" of the female. Although otherwise a relatively obscure figure, Dr. Johnston has been immortalized as the describer of this unique structure, now called "Johnston's organ." It remained for Louis Roth to elucidate more fully the role of the male's antennae. He found that males with their antennae removed or rendered immobile by placing a spot of shellac at their base were indifferent to flying females or to the sound of a tuning fork. Weighting the tips with shellac had a similar effect, but when the weighted tips were cut off, the males responded normally. After removal of all but a few of the feathery hairs on the antennae, males tended to ignore females, but often gave a mating response to the much louder sound of a tuning fork, and a few were able to direct their flight toward the tuning fork. Roth concluded that these hairs serve as sound detectors and that their vibrations are conveyed to the Johnston's organs at the base of the antenna; with two intact antennae the male is able to determine the direction from which the sound is coming and to act accordingly.

The male malaria mosquito (Anopheles) also responds to much the same tuning forks as the yellow-fever mosquito, although its mating posture is quite different: the male at first "hangs from the tail" of the female, then assumes a position facing away from her. The male Anopheles is able to depress the antennal hairs against the shaft, and during certain periods of sexual inactivity these hairs remain depressed. In contrast, the yellow-fever mosquito keeps his antennal fibrils constantly extended and is always ready to mate.

As Roth points out, these are basically "household" mosquitoes, and may be exceptional in that the sexual approach is direct and does not involve a preliminary courtship dance or "swarm," as occurs in most mosquitoes, gnats, and midges. Swarms of midges are well known to everyone, as they often form a column above one's head as one walks about in the fields. These swarms consist mostly or entirely of males,

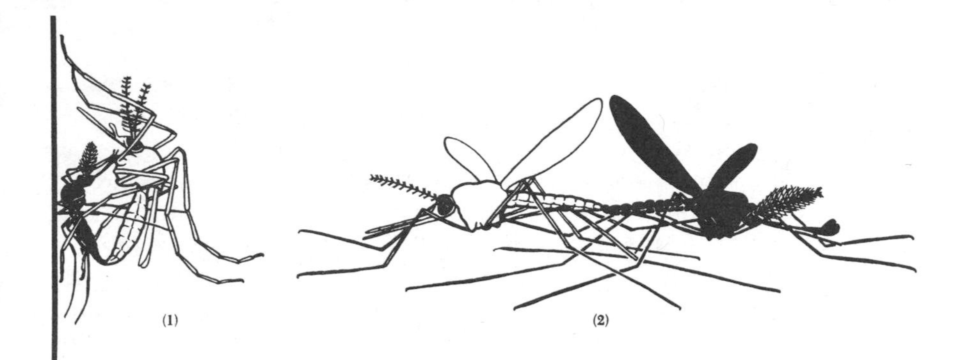

Mating posture of the yellow fever mosquito (*1*) contrasted with that of the malaria mosquito (*2*). In each case the male is shown as solid black. A good many insects besides malaria mosquitoes (cockroaches, for example) copulate while facing away from one another—an unsympathetic person might assume they can't bear to look at one another. (After Louis Roth.)

and it is often assumed that the sound of their wings plays a role in their assembly and perhaps in luring females into their midst. The swarming of midges has nowhere been described so well as by John Keats in his poem *To Autumn:*

> Where are the songs of Spring? Ay, where are they?
> Think not of them, thou hast thy music too,—
> While barred clouds bloom in the soft-dying day,
> And touch the stubble-plains with rosy hue;
> Then in a wailful choir the small gnats mourn
> Among the river shallows, borne aloft
> Or sinking as the light wind lives or dies . . .

The importance of sound to swarming midges or mosquitoes is not hard to demonstrate. Frederick Knab, a distinguished landscape-painter-turned-entomologist, commented regarding a swarm of mosquitoes milling above tufts of tall grass near Washington, D.C.: "The sound of the voice in conversation would, at the first shock, precipitate the entire swarm about a foot, and the mosquitoes would continue in rapid and confused movement while the speaking continued. . . . When silence was restored the swarm immediately resumed its normal flight. The experiment was repeated many times and each time the effect was instantaneous." The playing of a musical instrument to a dancing swarm can sometimes produce unusual effects, and a gunshot or loud whistle will often send them into wild disarray.

The fact is that the "wailful choir," though often characterized as producing a "courtship song," has never been shown to attract females into its midst. Furthermore, there is now reason to believe that the assembly of the males themselves is brought about not by sound but by a common response to certain landmarks. A single male will, in fact, form a private little "swarm" all by himself if no other males are present. In the words of J. A. Downes, of the Entomology Research Institute, Belleville, Canada, "no gregarious factor is necessarily involved in swarm formation but only the individual reactions of the insects to a common marker." Apparently the responsiveness of a swarm to sound is merely a by-product of its collective alertness to the sound of the female.

Swarms that form in a column above certain objects are often called "top swarms." Markers include such things as bushes, paths, pools, or even persons. As Harold Oldroyd notes, if two people walk together and then separate, their top swarms will often divide too. Midges and mosquitoes sometimes form swarms of such size and density over tree tops that the "plumes" they form can actually be photographed from a distance. Swarms over church steeples have, on more than one occasion, been interpreted as columns of smoke, resulting in a screech of sirens that sends the flies off in all directions as if in search of some great, unhappy female.

Dr. Downes has experimented with artificial swarm markers and in this way has elucidated the properties of markers to which the flies respond. For example, a small gnat that occurs naturally over pads of cow dung lying in a pasture will form swarms over a dark cloth, demonstrating that the odor of the cow dung is not the attractant. At the same time, a mosquito that Downes found along a sandy road developed swarms over a white cloth lying on a dark background. Such cloths could be moistened or could be treated with various chemicals, all with no effect on the flies, which evidently respond only to the image of the marker against a contrasting background. Different species of flies tend to respond to different kinds of markers or to fly at different heights above the same marker. For example, in Manitoba six species of mosquitoes, all of the same genus, often swarm at the same time. However, one of them forms swarms over clumps of caribou moss, another over the twigs of dead bushes, another between pairs of spruce trees. The remaining three occur at the margins of pools, one forming a swarm at a height of three to twelve feet slightly beyond the upwind margin of the marker, another at ten to fourteen feet directly above the marker, another at twenty to thirty feet above the marker. It appears that these swarms function to concentrate the populations of these species, which may have bred in scattered places in the area. Since the males will attempt to mate with any female producing a wing sound within a broad range of frequencies, the importance of gathering each species together into its own "club" is obvious. I am reminded of the way the scattered members of various religious sects in my community suddenly gather in their own churches and temples once a week—with most of the parents hoping, of course, that their offspring will eventually pair off with someone in their own sect.

Downes has shown that swarms of midges and mosquitoes begin when many individuals, in response to certain temperatures and light intensities, take wing and fly upwind until they position themselves over the windward margin of a swarm marker characteristic of that species. Then they allow themselves to be carried backward by the wind until they approach the leeward margin of the marker, whereupon they begin to control their flight again and finally once again fly upwind to the windward margin. With many individuals performing such movements at once, the effect is of a "dancing swarm." As one worker has remarked, the swarm seems to be attached to the marker "like a balloon tied to an object by a string."

In addition to top swarms, two other types of swarms sometimes occur. "Free swarms" occur over a flat surface devoid of prominent landmarks, and one assumes the flies respond to one another or to certain distant or subtle landmarks, though we do not really know. "Ceiling swarms" occur when great numbers of midges swarm rather high over a considerable area, forming a "ceiling" from which vertical columns may descend. Since these insects are most active during periods of fairly rapid changes in light intensity, most swarms of all types occur around sunset and sunrise. Mosquitoes kept under conditions of constant light intensity show no periodic activity cycles, indicating that swarming is not in response to an "internal clock," as some other rhythmic activities of animals appear to be. Swarms tend to form in the same place each evening and sunrise, and experiments with marked individuals show that the same mosquitoes, by and large, reform the swarm each time, even when there are several swarms of that species in the vicinity.

Many persons have assumed that mating occurs in the swarm, even though swarms are almost invariably found to consist entirely of males. Presumably females are seized as soon as they enter, and never really participate in the swarm. Some species are known to mate apart from the swarm or even during special "mating flights" that are independent of the swarm. Males of a New Zealand mosquito are said to land on floating pupae of females, slit them open with their "genital forceps," and mate with the females before they emerge, all the while fending off other males with their beaks. Behavior such as this has led some workers to propose that, at least in some cases, the swarm no longer serves as a "premating flight" but simply as a "ritual." It seems odd, to say the least, that midges and mosquitoes should expose themselves to swallows, swifts, and bats in such numbers simply to preserve a "ritual" when in fact their tiny brains are hardly capable of mollification by ritual.

For many years the leading opponent of the idea that swarms are closely related to mating has been Erik Tetens Nielsen, of Femmöller, Denmark, formerly with the Florida State Board of Health. Nielsen has done extensive research on the contents of swarms, using a great variety of techniques. He found it possible to "call down" certain swarms by singing two notes:

Higher swarms could not be called down, so he devised a net attached to a hydrogen-filled balloon. When the balloon reached the desired location, the net could be made to rise rapidly, taking in much of the swarm, and then close with a noose. In this way he was able to capture and study a great many swarms. He came to the conclusion that swarming

E. T. Nielsen's device for capturing swarms of midges. The cord is originally strung through a loop above the net; when the net reaches the desired location, a jerk of the cord opens the net. When the net is retracted quickly, it once again closes, hopefully containing a milling swarm of midges.

has little or nothing to do with mating. Rather, it appears to be, for the males, an "end in itself," perhaps functioning to "prevent inbreeding by mixing males from different breeding places." Nielsen points out that the fact that mating has, on occasion, been observed to occur in swarms does not necessarily mean a great deal. Most males are, after all, never seen except when they swarm. Furthermore, an occasional mating in a swarm does not necessarily imply that the swarm is a prerequisite for mating. "It is true," Nielsen comments, "that looking at the stars, to some young people, may be one of the first steps in an amorous affair, but it is unjustified to conclude from this that the study of astronomy is [pre-mating] behavior in *Homo sapiens*." I cannot resist quoting Nielsen's conclusions further:

"As to the question: Why do mosquitoes swarm? We have to reply: We do not know. It seems to us that the right thing to do is to admit our ignorance and abstain from any guesswork. . . . It is wonderful when we can piece together our bits of knowledge about the habits of insects and get them to make sense, but there is no reason to begin a panicky guessing-game because we, with our present ignorance, are unable to understand the purpose of some habit.

"In this situation we should remind ourselves of Steno's words . . . 'It is beautiful what we see, and still more beautiful what we understand. But by far the most beautiful is what we do not know.' "

Despite Nielsen's eloquence, there are still many who believe that swarms play an important role in bringing about the meeting of the two sexes of one species, at the same time preventing these insects—lacking as they are in sexual pheromones or high visual acuity—from wasting their energies trying to mate with members of an alien species. Downes has commented on the need for more actual counts of matings in swarms, and Nielsen, in a discussion of items of equipment useful in studies of this nature, has concluded that "no method of approach [can] compare in value with that of direct observation." Perhaps if fly-watching were to become fashionable, like bird-watching or the admiring of autumn foliage, we would solve some of these riddles.

There is evidence that both males and females of some gnats and mosquitoes congregate at the food source and that mating occurs here; in this case the food source serves as a "swarm marker." The females of certain midges feed on the pollen of flowers, and the males also gather here and "blunder into" the females. In some cases male mosquitoes are said to respond to the image of a warm-blooded animal and to mate with the females as they fly in to bite. Perhaps the most remarkable instance of this kind relates to a small gnat that mates only in the ears of jackrabbits; it is here that the females feed and are sought out by the males. Among the more "advance" kinds of flies, blowflies and the like, mating at the food sources and egg-laying sites (carcasses, in the case of blowflies) is very common.

Of special interest are certain flies that swarm only in unusual situations: smoke flies, for example. These flies are widely distributed but only very rarely encountered. However, in the 1920's and 1930's, several workers in Belgium and in England discovered that they gather in great numbers in the smoke of bonfires. In the 1940's, E. L. Kessel, of the University of San Francisco, undertook a search for one of our two apparently very rare North American species of this same group. After years of unsuccessful collecting around his home near San Francisco, Professor Kessel made a trip into Oregon and northern California, only to return empty-handed. Then, one August evening after his return, he found them flying over the chimney of his back-yard barbecue, in which he was burning weeds. The next day he went out and built a huge, smoky fire of weeds and green grass, and kept it going all day long. At first only a few smoke flies appeared, but gradually their numbers increased, until by early afternoon "they were dancing in great numbers in the thick of the smoke above the chimney, some strokes of the net yielding as many as thirty specimens. About a third of the flies captured at the time were females." In the course of the day Kessel collected about five hundred specimens, only a small part of the swarm but infinitely more specimens than had ever been collected previously. He did not actually observe mating in the swarms, and it remains to be learned why the flies swarm in a situation that occurs so irregularly in nature.

Having learned how to attract smoke flies, Dr. Kessel traveled widely in the West, north into Alaska and Yukon, and by building smoky fires was able to collect them nearly everywhere. In the course of this, he found that the flies would often run around on his clothes for hours or even days after the clothes had been exposed to smoke. Evidently the flies responded to the odor, absorbed by his clothing, and not to the sight or heat of a bonfire.

Other members of this same family of flies, inelegantly called the "flat-footed flies" because of the broad expansions on the hind legs of the males, have been found to swarm in the absence of smoke, chiefly in sunny openings in forests. Here the males dance in a cloud and dangle their hind legs, which resemble small, aluminum-colored flags sparkling in the sun. The females gather on foliage below the swarm and from time to time enter it, shortly to emerge in the embrace of a male. In this case short-range visual clues are apparently much more important than they are in the midges and mosquitoes, which rank a good deal lower on the evolutionary scale of the flies.

The use of visual signals, combined with wing sounds, is shown dramatically by the robber flies, which are large insects, and not the least bit secretive about their love lives. Many robber flies have patches of silvery hairs on their bodies or legs, and when a male is courting he displays these prominently while producing a loud hum that sometimes changes pitch and intensity in different phases of the courtship flight—about as close to a melody as insects can produce. Sooner or later the male makes a mad and noisy dash after the female, who may accept him or simply chase him away. In some species the female also has glistening patches of hairs which she is said to display to the male when she is "in the mood." I remember being entertained by robber flies on several hot Texas afternoons when my wasps were not doing anything nearly so exciting.

Robber flies are vicious predators on other insects. (Sometimes they are called "assassin flies," which seems a good deal more appropriate.) Now and then one finds a female feeding on a male of her own species, suggesting that a courtship went slightly awry. It is not surprising that in some robber flies the male normally courts and mates with females that are already feeding. It is also not surprising—to one who has learned to expect almost anything from insects—to discover that the males of a group of predaceous flies related to robber flies routinely capture insects and present them to the female prior to mating. These flies are called "dance flies" because of their tendency to form clouds of rapidly moving individuals over streams and in woodland glades. The swarms consist mostly of males, many of which carry a small insect that they have captured. When a female enters the swarm, she is attracted to a male holding a particularly desirable morsel, and the male is permitted to copulate with her as long as the morsel lasts. One observer noticed a male bearing two insects almost as large as he was. Since anticipation is too much to expect of a fly, we shall have to pass this off as an error of instinct.

There are hundreds of species of dance flies, and they exhibit many different kinds of courtship behavior, some of them so strange that they have been the source of much puzzlement and disbelief. The person who, more than any other, has elucidated these elaborate mating antics is A. H. Hamm, a self-educated British entomologist who devoted much of his life to the study of living insects in their natural habitat. Hamm found that male dance flies that merely carry an insect for the female to feed upon represent one of the simpler stages in a progressive series. In more advanced dance flies, the males wrap the prey in silken threads or sticky secretions that apparently prevent it from escaping. Still others spin strands of tiny bubbles from their anus to form a sphere in which the prey is embedded. These are the so-called "balloon flies." Some species have progressed to the point where the prey is very small and inconsequential, and is sucked dry first and plastered to the balloon in pieces. Evidently the prey still serves as a stimulus to the male to make the balloon, but the female appears to pay no attention to it; the balloon itself has become the stimulus for mating. E. L. Kessel, whom we met earlier as an authority on smoke flies, discovered a flight of these insects outside his window one day some years ago, and it was this discovery that stimulated him to begin a study of dance flies. In his words:

"One morning, while looking out of our kitchen window, I observed a number of conspicuous white objects scintillating in the morning sunshine as they zigzagged back and forth close by a Monterey pine. They were flying at an altitude of some fifteen or twenty feet, and it was only by tying my net handle to a sturdy surf-casting rod and then mounting to the top of a six-foot stepladder that I was able to bring any of the specimens to net. They proved to be all males and the white object which each carried was a delicate, balloon-like structure which invariably had a minute [insect] plastered into its anterior surface. . . . These minute captured insects are always oriented so that the head, and often the thorax as well, projects free from the balloon, being directed forward and upward. And this projecting part of the prey's body is used as a handle which the male always seems to have hold of with one or more of his feet. . . .

"During the first years of my study I was able to observe the mating activities of this species. On several occasions I saw a female join the dancing group of males and select a mate. As the two embraced they lost altitude for a moment, and then floated off together to settle on nearby vegetation. Examination of the paired flies revealed that the balloon had been transferred to the female. In the act of mating the male hangs from the vegetation by his front feet, his middle legs supporting the female's thorax, and his hind legs holding her abdomen. The female holds on to the balloon with all of her feet; never does she grasp the prey as a handle for the balloon in the manner of the male. Instead, she keeps turning the balloon from one position to another during the mating period, apparently entirely unconcerned with the prey and certainly making no attempt to feed on it."

As for the ultimate stage in evolution: you have already guessed it. Of course, the male no longer captures prey to present to the female, but merely offers her a balloon spun of anal secretions. The female fondles this with her legs during mating, her predatory urges apparently completely forgotten. It is as though our own giving at weddings and Christmas had reached the point of our putting up beautifully wrapped empty boxes—demonstrating a nobility of sentiment, perhaps, but not at all good for the economy.

Unfortunately, the first balloon flies to be discovered were these very specialized species, and as can be imagined they left their discoverers somewhat bewildered, lacking as they did knowledge of the several intermediate stages. It was in 1875 that Karl Robert Romanovich, Baron Osten Sacken, discovered the first balloon flies. While vacationing near Berne, Switzerland, Osten Sacken discovered a swarm of small flies dancing in sunlight penetrating a fir forest. "What attracted my attention to them," he wrote, "was the uncommonly brilliant white or silvery reflection which they gave in crossing the sunbeam. I caught one of them with my forceps, and was astonished to find a much smaller fly than I had suspected, and without anything silvery about it. . . . However, I perceived . . . not far from the fly, a flake of opaque, white, film-like substance . . . so light that the faintest breath of air would lift it. . . . What is the purpose of this performance? . . . Where do they obtain these flakes?"

Baron Osten Sacken, by the way, was a professional diplomat, born in Russia, and for several years secretary of the Russian Legation in Washington, later consul general of Russia in New York. But he had "flies in his blood," having published his first paper on them when he was in his twenties. While engaged in "diplomacy," he amassed a great collection of flies, and published extensively, including the first catalogue of North American flies. He retired at forty-three to devote all his energies to the study of flies, and although he eventually settled in Germany, he not only left his large collection of American flies at the Museum of Comparative Zoology at Harvard University, but with some difficulties persuaded the prolific German fly specialist Hermann Loew to do likewise.

We can forgive Osten Sacken his bewilderment as to the source and function of the "balloons" he had discovered. At least he refrained from theorizing; in his autobiography he wrote that he believed "all phenomena of life . . . to be susceptible of being described, but not explained." But others less restrained soon discovered balloon flies or read about them. One suggested that the balloons were aerial surfboards enabling the flies to glide about on the sunbeams, another that they were used to warn off predators, still another that they were used to draw the attention of the female—which is fairly close to the truth. No sensible theory as to the origin and function of this unusual behavior was, of course, possible until the discovery some years later of several intermediate stages connecting the builders of balloons to simple predators. We now recognize this as a wonderful example of how comparison of related animals can provide clues as to how unusual structures and behavior patterns may have arisen.

One of the modes of courtship behavior described by A. H. Hamm is even more unusual than those recounted above, and perhaps represents the final stage in a separate evolutionary series. One day, while observing quietly along a stream, Hamm noticed male dance flies picking up various bright objects from the surface, usually insects but sometimes pieces of leaves or flowers. He threw petals of buttercups and daisies into the water, and the flies picked them up and carried them about like banners, eventually transferring them to the females during mating. One is reminded of one of Louis Agassiz's classic remarks: "The possibilities of existence run so deeply into the extravagant that there is scarcely any conception too extraordinary for Nature to realize."

Most of the "higher" flies—that is, more robust flies of the general form of the housefly, which exhibit many structural advances as compared to midges and which have evolved in fairly recent geologic time—do not swarm in the manner of midges and dance flies. The few that do fly in swarms usually maintain position not by flying into the wind and drifting backward, as described for midges, but by hovering for long periods in one spot. It is not uncommon to see such flies hovering in a diffuse cloud beneath a tree or in an opening among trees. The British entomologist O. W. Richards, for example, writes of one of the flower flies:

"I have seen the two sexes hovering in the air opposite to one another, as if they were each suspended from a string; at regular intervals they knocked their heads together and then swung apart again. As they did this they slowly sank to the ground, the male maintaining the whole time a shrill hum. . . ."

Male horseflies have been observed hovering in openings among trees, each remaining in one spot and maintaining a certain distance from his neighbors. They dash after other insects that enter their field of vision, and drive away other males that approach too closely; sometimes a male takes off in pursuit of another, and still others join in, forming a "chain reaction" that may involve most of the swarm. When a female enters the swarm she is apparently seized by one of the first males whose territory she crosses, and the two drift out of the swarm and onto vegetation, where mating is completed.

However, the majority of higher flies have abandoned such premating swarms in favor of more intimate kinds of courtship, although some do establish territories on the food source. For example, males of some of the walnut-infesting fruit flies alight on husks that are in a condition suitable for egg-laying, often remaining there for hours and chasing away any other males that arrive. It is said that the resident male lunges at the newcomer, waves his wings, then, if still unsuccessful in driving him off, stands up on his hind legs and strikes at his opponent with his front legs. After a short "boxing match" the intruder is usually willing to look for another nut. If, however, the intruder is a female, the resident male becomes greatly excited, dances back and forth, whirls about, and fans his brightly patterned wings. When the female settles down to egg-laying, the male stands somewhat to one side or circles her. As soon as egg-laying is completed, the male mounts the female, who spreads her wings and extrudes her genital organs. Mating lasts several minutes, and the female may then lay more eggs, mate again, lay more eggs, and so on. Finally she departs, leaving the male to patrol the nut until another fly arrives.

An even more remarkable courtship behavior was described many years ago by William Morton Wheeler of Harvard. The fly was a tropical "stilt-legged" fly, not unrelated to the fruit flies. I can certainly not improve on Wheeler's colorful prose:

"A dozen or more individuals of both sexes select and for many days use as a playground the large horizontal leaves of some bush growing along the edge of the jungle trails. . . . The flies run about on the upper surfaces of the leaves with rather jerky movements or flit from one leaf to another. They evidently recognize one another by sight. The females are very coy and often drive the males away by making sudden lunges at them when they approach too closely. The males, too, occasionally fight with other males, rising perpendicularly on their long hind legs and facing one another. . . . If carefully observed when alone, the male is sometimes seen to stand still and regurgitate from the tip of his proboscis a small drop of liquid, which he at once swallows, only to produce another drop and withdraw it in turn. This may be repeated several times and is obviously the same as the behavior described . . . in the common housefly. In both cases the alimentary canal is distended with liquid and the indecent creatures amuse themselves by alternately regurgitating and swallowing portions of their food.

"If the female . . . is willing to receive the male, he is permitted to approach within a few centimeters. Facing her he then performs a peculiar dance, stepping first to one side and then to the other, swaying his abdomen towards her and at the same time downward till it strikes the surface of the leaf. . . . Perhaps [in this way he is] displaying to its best advantage the beautifully iridescent surface of his terminal abdominal segments. At any rate, the female, after witnessing this *danse du ventre*, seems to indicate to the male by some very subtle sign, which I have been unable to detect, that she is ready to receive his embrace. She instantly bends her body in an arc by throwing her head back and turning up the tip of her abdomen. The male is on top of her at once and bringing his proboscis in contact with hers, places a drop of food on it and almost at the same instant inserts his intromittent organ into the tapering apex of her abdomen. The female quickly straightens her body and the pair now remain together usually for as many as ten to fifteen minutes. . . . Occasionally [the male] reaches forward with his fore tarsi and scratches the female's eyes, but more frequently he advances his head and proboscis and with a peculiar pecking movement places a minute drop of regurgitated liquid on the upper corner of one of her eyes. She at once reaches up with her forelegs, wipes off the droplet with her tarsi and draws them over the proboscis. . . . After mating the female may in a short time accept another partner. When she does this she is, perhaps, actuated more by hunger than by lust."

The fly has obviously come a long way from the indiscriminate responses and clumsy approaches of the mosquito: a dolt become a Don Juan. In many species of higher flies, the body and wings are patterned so as to enhance short-range visual recognition of the species and sex and to provide "sign stimuli" that facilitate mating. Wing flicks, body turns, leg movements, licking of body parts, feeding of the female by the male, and a variety of other intimacies occur. Sounds are usually less important than in the mosquito, if they occur at all, but some exceptions are known. For example, the Queensland fruit-fly male is said to "call" the female toward sunset by standing stiff-legged and vibrating his wings rapidly. The call consists of "high, flute-like notes emitted in series, each note varying in duration from a half to two seconds." There may also be an exchange of odor stimuli, usually over very short distances. However, some female fruit flies apparently produce a powerful pheromone that attracts males from a distance. Oil of citronella is a powerful attractant for some male fruit flies (even though it is a commonly used repellant for mosquitoes and gnats). In this case, the citronella is believed accidentally to resemble the sex pheromone of the female. It is probably for the same reason that male Mediterranean fruit flies are attracted to kerosene.

The little fruit-fly Drosophila breeds rapidly in small bottles—which of course is one reason why it is such a popular laboratory animal. Condemned as it is to carry out its private life in a glass bottle, Drosophila has few secrets. But then, who does these days, with a treatise called *Human Sexual Response* on the best-seller list? In the case of Drosophila, males and females respond to the sight of one another's motions, and some species refuse to mate in the dark. However, courtship and mating depend mostly upon interchanges of odor and touch stimuli. The male initiates courtship by tapping the female with his front legs. This is apparently his primary means of being sure he is confronting a female of his own species; apparently he "tastes" her by way of sense organs at the tips of his legs, just as a housefly is able to taste food with its feet. If he finds he has encountered a female, he stands facing her and brings out one wing at a right angle to his body and vibrates it rapidly for a few seconds. He then circles about, facing the female, and from time to time repeats the wing fanning, always using the wing closest to the female. Sooner or later he dashes up to her and licks her genital organs. At this stage the female is able to detect whether or not he is a male of her own species, apparently chiefly by odor, since females with their antennae cut off do poorly at telling males of their own species from those of others. A receptive female spreads her wings and genital organs, and the male mounts, often stroking the female with his forelegs during mating.

This generalized description makes it sound as if courtship were simple and stereotyped, which in fact it is. However, we now know that each of the many species of Drosophila has its own particular repertory of movements. A detailed comparison of the mating behavior of 101 species of Drosophila was made some years ago by Herman T. Spieth, then of the City College of New York, now a professor at the University of California at Davis. Closely related species were often found to have very similar mating postures, but in these species differences in sexual odors apparently serve to prevent interbreeding. Species more unlike structurally tend to have very different courtships, largely confirming the classification concocted from dead specimens in museums.

The mating of houseflies, bluebottle flies, and their kin generally takes place on the ground or on a source of food: a roadside picnic area is often a good place to observe them. In the words of the authors of a recent research paper on the mating of the housefly: "Watching the mating behavior of this ubiquitous insect is undoubtedly a human pastime of great antiquity." Aristotle is said to have been one of the earlier students. Even Shakespeare's mad King Lear, wandering in the fields near Dover, was moved to comment on flies and men:

> The wren goes to 't, and the small gilded fly
> Does lecher in my sight.
> Let copulation thrive; for Gloucester's bastard son
> Was kinder to his father than my daughters
> Got 'tween the lawful sheets.

The male housefly is a restless animal, forever ready to mate with a female, another male, or even a dead fly. However, a live female definitely has the most attraction, and when he approaches one he pauses briefly before leaping abruptly and accurately upon her back. He then reaches forward and strokes her head with his front legs, usually all that is needed to charm the female into submission.

A team of workers headed by William Rogoff, of the United States Department of Agriculture laboratories at Corvallis, Oregon, has recently asked if the female fly produces a pheromone that attracts the male to her presence. They found that males definitely prefer a current of air passed over frozen females to one passed over frozen males—they used frozen females not to test the males' ardor, but simply as a way of ruling out the movements of the female or any sounds she might make as factors in attraction. They also prepared "pseudoflies" by cutting off bits of black shoestring and impregnating them with "essence of female fly," the latter being made by grinding females in benzene, pouring off the soluble part, and drying the remainder. The males jumped upon these female pseudoflies quite readily, and much preferred them to male pseudoflies, although in fact they had a greater tendency to jump on male pseudoflies when female pseudoflies were nearby, demonstrating their greater excitability when "essence of female" was present.

The mating of houseflies may be slightly less sophisticated than that of dance flies and stilt-legged flies, but there is no doubt that it gets results. The fertilized female housefly lays several hundred eggs, and these eggs hatch in only eight to twelve hours, giving rise to maggots that grow very rapidly and produce another crop of adult flies in only a week or two. L. O. Howard, for many years chief entomologist of the United States Department of Agriculture, and author of one of several books devoted exclusively to the housefly, once attempted to calculate the number of flies that might result from a single initial fertilized female in the course of an average summer in Washington, D.C., assuming that all survived. His hypothetical housefly laid her eggs on April 15, and by September 10—well before the end of the fly season in Washington—her progeny numbered 5,598,720,000,000. Another writer concluded that the offspring of a single pair of flies in the course of a summer would cover the

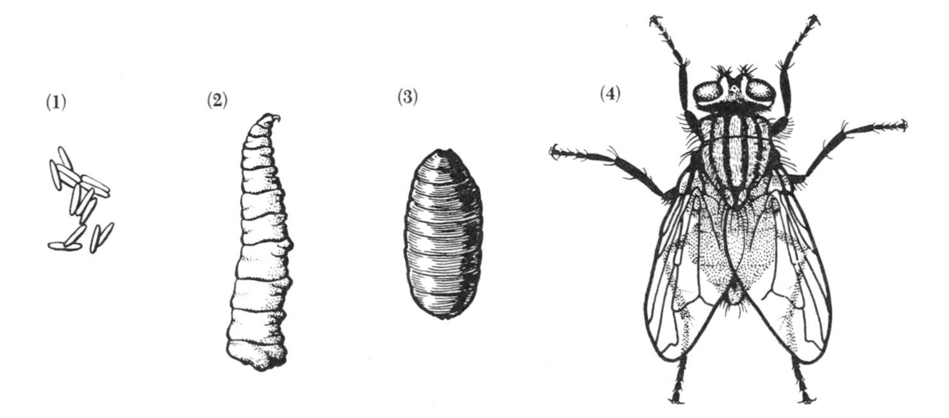

Life stages of the housefly. The eggs (*1*) are laid in masses in excrement or rotting substances, producing maggots which have no legs and no visible head save for a pair of hooks at the tapered end (*2*). The fully grown maggot forms a pupa inside the hardened larval skin (*3*), producing a fly all too familiar to everyone (*4*).

earth to a depth of forty-seven feet, if all survived. Harold Oldroyd checked these figures and found that "a layer of such thickness would cover only an area the size of Germany, but that is still a lot of flies."

The housefly, like many of its relatives, may be classified as a "filth fly," which means simply that it breeds in things we consider foul and messy—though the fly doubtless classifies them as delectables. Horse droppings are special favorites, although human excrement is also popular, and Dr. Howard lists as other breeding places pig and chicken manure, rotting vegetables of several kinds, "kitchen slops," rotting chicken feathers, and refuse from slaughterhouses. These materials are so diverse, and their nutritional value so limited, that some people have wondered exactly what it is that the housefly maggots utilize as food. Z. H. Levinson, of the Hebrew University, Jerusalem, recently discovered that when small housefly maggots were introduced into a wheat-bran-milk mixture that had been sterilized, they failed to develop. But when the mixture was inoculated with bacteria common in the digestive tracts of man and his domestic animals, they began to grow, and when the concentration of bacteria reached a very high level the maggots developed rapidly. Dr. Levinson concluded that "the diverse materials on which housefly larvae seem to thrive in Nature are merely substrates for bacteria, and these are most likely to be the actual food of the larvae."

This may well be true of many of the filth flies, including the blowfly maggots that were once used in medicine for cleaning out wounds. There is no question that many of the higher flies play an important role in the disintegration of carcasses, fecal material, rotting plants, and so forth. Flies are often attracted to such materials as soon as they are available. We know that some of them are attracted from considerable distances by the odors of skatole, indole, ammonia, and other odorous substances present in feces and decomposing materials. Different species of flies tend to have slightly different preferences as to type of materials, amount of moisture, state of decomposition, and so forth. In a given situation, there may be a succession of different flies and other insects breeding, one following another as the material decomposes or dries out.

Some years ago Carl O. Mohr, then a student at the University of Illinois, but now at the University of California at Berkeley, made a study of the succession of insects occurring in cattle droppings in Illinois pastures. He found that the first flies to appear were horn flies, blood feeders that are constant attendants of cattle and that go through their larval stage in fresh fecal material. Adult horn flies lay their eggs quickly, whenever fresh droppings are available, and leave the droppings promptly. Mohr quotes C. V. Riley, who observed an animal in the act of defecating:

"As the operation commenced, forty or fifty flies moved from the flank to the back of the thigh . . . and at the close of the operation they were seen to dart instantly to the dung and to move quickly over its surface, stopping but an instant to deposit an egg. . . . Every individual had returned to the cow . . . in little more than a minute."

Within a few minutes several kinds of filth flies, comparable in size to the housefly and the horn fly, make their appearance and depart after laying their eggs. Over the first few days the maggots of these first arrivals develop rapidly in the dropping. Some time within the first few days several species of small "dung flies" make their appearance, and often remain in attendance for several days, their larvae developing in the dung mostly after the first crop of maggots has finished its development and left. Some of these dung flies are very abundant creatures, walking incessantly over the dung and constantly fanning their wings, which often have a small dark spot at their tips. Sooner or later various dung beetles make their appearance, as well as a variety of flies, wasps, and beetles that are parasitic or predatory on the dung feeders. As the dropping becomes increasingly dry the insects decline in numbers, and of course parts of the dung may be balled up by dung beetles and buried beneath the dropping or rolled away and buried elsewhere. Naturally, the succession of insects differs at different seasons and in different areas, and is greatly influenced by the amount of light and heat the dropping receives, the amount of rainfall, and so forth.

Other workers have studied the sequence of insects occurring in carcasses. In their studies of the filth flies of Guam, conducted at the close of World War II, G. E. Bohart and J. L. Gressitt, who have both since gone on to distinguished careers in other aspects of entomology, took occasion to make regular observations for about a month on a human corpse lying on the beach. When they first found the corpse it was about four days old, and great numbers of maggots of the big-headed blowfly were already leaving the corpse to form their puparia in the soil nearby. At this time a second species, the red-marked blowfly, had made its appearance, and the females were laying eggs in the decomposing flesh. A week later "the corpse had lost its bloated appearance and the flesh was easily pulled from the bones"; at this time great numbers of blowflies of several species swarmed in the nearby vegetation, and maggots of the red-marked blowfly were leaving the body, although other maggots were still developing in it. Two weeks later the corpse had dried out considerably and had a musty smell, and the major inhabitants were various beetles, although larvae of several kinds of small flies were present in parts of the carcass and in the sand underneath.

Information of this kind can be of considerable value in estimating the time of death of a human body, a matter sometimes of major importance in criminal investigations. Many years ago Pierre Megnin made a study of the succession of insects in buried corpses in France, from which he concluded that one can often determine the age of a corpse fairly accurately by this means. In 1898, M. G. Motter assured his own immortality by reporting on the animals present in 150 disinterred bodies in Washington D.C. His paper, titled "A Contribution to the Study of the Fauna of the Grave," is fascinating reading, and tells one more about his own future than he is commonly rewarded with. The most common insects are springtails, various "rove" beetles (very common inhabitants of carcasses in advanced stages of decay), and coffin flies. The latter are minute, bristly flies that apparently crawl through soil readily and pass through many generations in buried carcasses. It is difficult to see any pattern in Motter's results; recently buried bodies often had much the same fauna as very old ones, except that corpses buried for more than thirty years were often so thoroughly decomposed that they had little in the way of a special fauna. Dr. Motter found that so many variables were operating—time of year of burial, age of person at time of death, cause of death, nature of soil, type of coffin, and so forth—that few generalizations were possible. Most of all, he was impressed with how little we know about the life histories of the insects involved—and I am not aware that things have improved much in the past seventy years. Motter took issue with Megnin's conclusions, and did not feel that legal evidence of any importance could be obtained from a study of the insects in a buried corpse. Nowadays our morticians insist that we buy expensive, hermetically sealed caskets, so I suppose we may look forward to a much longer period of decomposition. Doubtless the hermetically sealed casket, like the jukebox, will go down in history as one of the crowning achievements of Western civilization.

Flies being what they are, it is not surprising that some have taken to feeding on living flesh rather than on carrion. One of the best known of these is the famous "screwworm," a major pest of cattle and other domestic and wild animals in warmer parts of the globe. The adult is a blue-green blowfly not very different in appearance from many of the filth feeders, but she lays her eggs on the margins of wounds in the skin, such as brand marks or barbwire scratches, and the maggots feed in the wound and soon invade sound tissue around it. The animal becomes ill and eats little, and in the meantime the wound grows larger and attracts still more flies, finally, if untreated, causing the animal's death.

Screwworms and other flesh feeders do not hesitate to attack man, particularly if he is wounded or has running sores, though usually they are not allowed to develop to a dangerous level. Infection with fly maggots is termed "myiasis." In his monograph *The Flies That Cause Myiasis in Man,* Maurice James, of Washington State University, describes the effects of various maggots living in the human body. In certain parts of the world it is dangerous to sleep outdoors in the daytime, especially if one has nasal catarrh, as flies may enter the nose and deposit larvae. In such cases the nose and face may become greatly swollen, and severe headaches occur. "The breath becomes bad, and a discharge consisting of a mixture of pus and blood is passed through the nose. The patient is in intense pain, more severe at times, but more or less constant. . . . The septum of the nose may fall in, the soft and hard palate may be pierced, the pharynx may be eaten away to the bone, and even the hyoid bone may be destroyed." The maggots of some of the myiasis producers are remarkably tough. Professor James reports that some have been kept in 95 per cent alcohol for an hour and have been known to produce adults; they will also survive for some time in hydrochloric acid, turpentine, or carbolic acid.

Maggot infestations may also occur at other openings in the body: the eyes, ears, anus, or urogenital tracts. In some instances more than a quart of maggots have been removed from patients. A wide variety of fly larvae, including some otherwise relatively harmless ones, are able to establish infections in the digestive tract if swallowed. In this case one is likely to void or vomit maggots sooner or later, an experience nearly as traumatic as the internal effects of the maggots.

In all these cases infection of man is somewhat accidental, but there are a few fly larvae that feed primarily on the flesh or blood of man. One is the Congo floor maggot, which lives in the mud floors of native huts and comes out at night to feed on the blood of sleeping persons. Another is the human botfly of tropical America, the larva of which lives beneath the skin of humans in much the same way that the botflies of cattle and other animals live. The female fly has the incredible—but well-authenticated—practice of laying her eggs not on a person but on a biting fly which she captures, usually a mosquito. In a few days the eggs hatch, but the larvae remain within their shells until the mosquito bites a person, whereupon they emerge quickly and either crawl through the small hole made by the mosquito or actually bore into the skin.

Some years ago Lawrence Dunn, of the Gorgas Memorial Laboratory in Panama, had the fortitude to work out the life history of the human botfly by rearing them on himself. Having discovered a fly to which eggs of the botfly had been attached, he allowed two of the botfly larvae to penetrate his left arm. A day later they showed as small red welts, which when viewed with a lens could be seen to have an opening to the air and to contain a tiny maggot that was forcing out small amounts of fluid. Much to his surprise, Dunn found that he had accidentally picked up four more bots, two on his right arm and two on his right leg. He let them all develop, even though each of them periodically itched severely, especially at night, and from time to time each produced sharp pains. The welts developed into large, open boils, which discharged so much pus, blood, and fluid that they had to be kept loosely bandaged. After forty-six to fifty-five days, the maggots dropped out, and each was placed in a test tube and an adult fly reared from it. The fully grown larvae were about an inch long and about a third of an inch in diameter. Dunn found that the wounds healed rapidly after the larvae had left, although one scar was still in evidence after eight months.

Myiasis is rare in the affluent, air-conditioned world of today's middle class, and in fact flies of all kinds are rarely more than temporary annoyances. Malaria, yellow fever, filariasis, sleeping sickness—fly-borne diseases that once swayed empires—have all been contained; the screw-worm is being sterilized; and the apple maggot and the cheese skipper have been banished from the A&P. Farmers nonetheless pay several million dollars a year to control crop-infesting flies, and local outbreaks of encephalitis, a mosquito-borne disease with a high rate of mortality, are reported in the United States every summer. Despite the discovery of atabrine and other new antimalarial drugs, over half a million United States servicemen contracted malaria in World War II. Since the war, we have heard a great deal about the eradication of malaria, and the World Health Organization has made some real progress in this direction. But at the same time mosquitoes are developing strains resistant to DDT; and a strain of the malarial parasite resistant to chloroquinine, the most effective drug yet devised, has disabled several thousand American servicemen in Vietnam for periods averaging thirty-five days each. We have by no means licked the housefly problem, for strains of houseflies resistant to various insecticides are now widely prevalent. Just in the last few years a very serious pest of cattle, the face fly, has become widespread in the United States. We have never been able to do much about sand flies or black flies because of difficulties in getting at their breeding places. In many parts of the world man and his domestic animals are almost constantly nagged by one fly or another, and without constant vigilance his fruits and vegetables become wormy. It would seem as though it would not take a very great decline in the world's standard of living—always a possibility in these precarious times—to cause us to capitulate to the flies.

With luck this will not happen, and with persistence we shall probably manage to solve most of our more pressing fly problems (and put a good many entomologists to work in the process). Harold Oldroyd, in the final chapter of *The Natural History of Flies,* writes with moderate optimism about the possibility of eliminating many of the blood-sucking flies. But he feels that the higher flies may have better prospects:

"They have learned to use decaying, fermenting, or putrefying organic materials, universal media that will always exist. No doubt we shall continue to campaign against the house-fly, but we shall not defeat it by chemicals, because it evolves resistant strains too quickly for us. Hygiene will keep it at bay in superior districts, but there will always be plenty of breeding material left for it. As quickly as urban areas are denied to the house-fly the tourist and his motor-car make more rural areas attractive to it."

In the last two decades much interest has attached to the possibilities of controlling flies and other insects by releasing great numbers of laboratory-reared males that have been sterilized by irradiation. In the 1950's, E. F. Knipling and his associates in the Entomology Research Division, United States Department of Agriculture, undertook such an experiment with the screwworm fly on the island of Curaçao, and in a relatively short time totally eradicated the pest. At the beginning of the experiment nearly all egg masses were completely fertile, but as they began releasing great numbers of sterile males from airplanes, successive generations of females were increasingly exposed to these sterile males, and consequently more and more of their egg masses were sterile. Finally, in only four fly generations, all were sterile. Knipling and his associates then turned their attentions to Florida, the only place in the southeastern United States where the flies overwinter successfully, although each summer they migrate northward in considerable numbers from this center. They bred more than 50 million screwworm flies each week in the laboratory, using more than forty tons of ground whale and horse meat to feed the larvae. More than two billion sterile flies were released from planes, and within a year the pest had been eradicated from the Southeast. The astonishing success of a method of control that "enlists the reproductive process of the species in its own extinction" earned Dr. Knipling the National Medal of Science in December, 1966. Similar techniques are now being applied to a variety of insect pests the world over.

There have already been some failures, and we now know that the screwworm fly was an exceptionally suitable insect for this kind of control. Not many insects can be so readily reared in huge numbers, and not all will mate normally after sterilization. Furthermore, eradication in broad, continental situations is quite a different matter from eradication in restricted areas such as Curaçao or peninsular Florida. There may also be problems in flooding the countryside with great numbers of pest insects, sterile and temporary though they may be. Experiments are now being conducted on chemical sterilants that can simply be applied in the manner of insecticides. As Knipling points out, such sterilants could theoretically be developed for use on rats, starlings, or any kind of animal we decided we did not want. It is now well established that the rendering of a large number of individuals sterile, but otherwise sexually competitive, has, in Knipling's words "a greater influence in reducing the [reproductive] potential of the population than does the elimination of the same number of individuals by destruction or removal." One hopes that such chemical sterilants will prove highly selective and that they will be used with the utmost caution.

There is a real possibility that intensive application of such sterilants will select out strains of insects capable of reproducing in spite of them, just as strains resistant to various insecticides have developed. It is also possible that as man's own population explosion continues he will amass such quantities of trash and wastes that a population explosion of those resistant and adaptable lovers of filth, the flies, will reach uncontrollable proportions. Whatever happens, we can be fairly sure that we are some distance from that "chemically sterile, insect-free world" feared by Rachel Carson. The flies will be with us for a long while, and we may as well learn to understand and appreciate them. Vincent Dethier, in his delightful book *To Know a Fly*, makes a strong case for the blowfly. Personally, I hope to seek out the coffin fly; when he finds me, it will be too late to enjoy him.

I particularly like the quotation from Lewis Carroll's *Through the Looking Glass* with which Dethier prefaces the first chapter of *To Know a Fly:*

" 'What sort of insects do you rejoice in, where *you* come from?' the Gnat inquired.

" 'I don't rejoice in insects at all,' Alice explained . . ."

Perhaps flies are a bit too dissolute, a bit too sinister to be a cause of rejoicing, but at least we can meet them halfway and get to know them well enough to enjoy their many remarkable traits. Here in America, north of Mexico, we have 16,130 species of flies to choose from, according to a recent catalogue compiled by workers in the United States Department of Agriculture. Unfortunately, cataloguing is only the first step to knowledge; once a species has been named and listed, we should start from there and try to learn something about it. There is nothing quite like a fly; and no fly is quite like another. The catalogue of flies is filled with comments such as "many vexing . . . problems remain unsolved" and "very little is known of the habits or immature stages." When we consider the many ways in which flies impinge upon man, and the remarkable life histories of so many of them, it is not much of a tribute to man's intellect that while reaching for the stars he ignores the fly on his windowpane.

Bedbugs, Cone-nosed Bugs, and Other Cuddly Animals

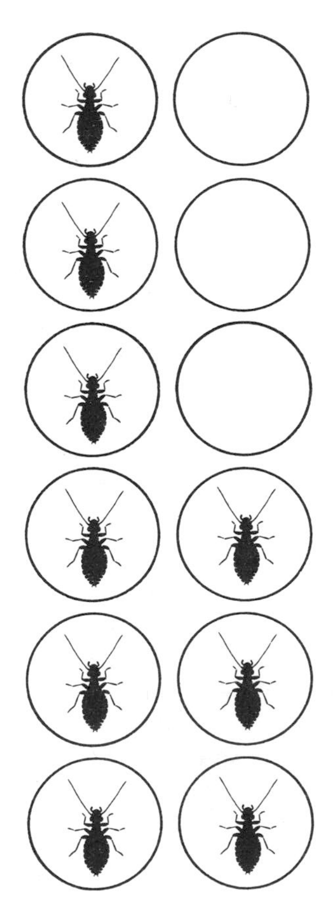

In 1889, the head of the United States Bureau of Entomology began an article with the remark that he had "occasionally met with a favored individual who had never seen a bed bug. But such fortunate people are rare. . . ." Times have changed. In our affluent, sanitized world the bedbug is but a feeble voice from the past, an almost mythical creature. I can vouch for the fact that he is more than a myth, as I spent one wild night about twenty-five years ago providing the drinks at a bedbug bacchanal. I was then a college student on my first trip to Florida, and my budget was such that each day I had a choice as to whether I would eat a good meal or sleep in a bed that night. That cabin in Lake Worth proved a poor investment of a dollar; I had a bed I didn't sleep in, and was myself eaten—I dare not say drunk, for fear you will misunderstand me. But I was already enough of an entomologist to make the best of it. I collected so many bedbugs that years later I was still using them in my classes.

I have never renewed acquaintance with the bedbug (knowingly, at least). And I am not about to launch an appeal for the preservation of bedbugs. If we must endure the loss of the bluebird and the decline of the butterfly, let us at least rejoice in the demise of the bedbug. He is not necessarily gone everywhere and for good, but at least he is in eclipse in our snug, smug little Western world.

Some idea of the former prevalence of bedbugs may be gleaned from the following quotation from another member of the United States Bureau of Entomology, a man with the exotic name Alexandre Arsène Girault (he was born in Annapolis, Maryland):

"On the night of October 29th, 1907, I arrived in Cincinnati, Ohio, near midnight and obtained a room at what is considered one of the largest and best hotels there. This room was on the second floor, and proved to be a rather small one, about 18 feet long and about 12 feet wide. It was elegantly and neatly furnished, with the walls painted a dark gray and ornamented with mural paintings of flowers; the floor was well carpeted. . . . [As I] approached the bed [I] observed a single [bedbug] resting on the spread. . . . I killed it. After this, I looked the bed over, and finally decided not to get into it, but to lie across it after disrobing, leave the lights on and obtain such sleep as possible under the circumstances. In carrying this out, I did not disturb the bed linen. After lying stretched out across the bed in this manner for about half an hour, I awoke, and upon looking around me, observed [bedbugs] rapidly crawling away over the bedspread, all of them swollen with blood and having the appearance of being very recently fed. . . . The time was about 1:20 A.M. Between this hour and 3:30 A.M., I dozed off from time to time, lying in the same place, but distinctly remember waking at 2 A.M. and 3:20 A.M. and discovering numerous specimens hurrying away over the coverlid. Each time I arose and killed all the bugs in sight, and also those which having been glutted from the host, had left it, crawled 2 or 3 feet away, and were hiding under the bedlinen. . . . At about 3:30 A.M. I decided to leave the bed, and passed the rest of the morning dozing away in a rocking chair."

Girault was dispassionate enough to refer to himself as "the host" and to gather a few notes on the bugs' reactions to light and their behavior after engorgement and to publish these as a scientific paper. A few years earlier he had published a study of the life history of the bedbug, using his own arms and fingers to feed the bugs periodically. A very patient and curious man, Girault discovered that bedbugs need to feed only once between each molt but that each meal they "gorge themselves until unable to hold more." The first-stage bug, before the first molt, feeds, on the average, for about three minutes, the second stage five minutes, the third stage six minutes, the fourth stage eight minutes, the fifth stage ten minutes, and the final (adult) stage ten to fifteen minutes (check Girault's figures for yourself if you are skeptical). Between each blood meal the bug hides in a crack for seven to ten days to digest its food. In another paper, Girault showed that bedbugs feed readily on mice, and can thus maintain themselves in places where no humans are readily available. It has long been known that bedbugs are able to live several months without feeding at all. Like any other animal, the hungrier the bug, the bolder he is.

Girault went on to a career in entomology that was something less than distinguished. Although employed by the Bureau of Entomology for several years as a specialist on certain small wasps, he was an individualist to the point of being unendurable to his colleagues. He eventually resigned, bought his own printing press, and began to print his own papers, spicing them with comments no editor would endure, such as "[my papers] contain facts which . . . cannot be ignored by atheistical or mechanistical or any other kind of science"; and "intellectuality pays off in everything but pay." His papers often concluded with a poem, of which the following is a prize example:

THE ENTOMOLOGIST

He's a man to whom the insect's all,
Whether Emperor moth or lowly flea;
Whether wings its way or can but crawl.
Is this the time you bid me come
To feast or dance?
Sir, much, much earlier with the ants
Was I at home.

Girault spent his later years in Australia, where he turned out hundreds of short descriptions of new parasitic wasps from that country—and a few from other, rather remarkable localities. Note the following new species, named after J. F. Illingworth, a colleague whom Girault felt had sold his scientific soul for financial reward:

"Shillingsworthia shillingsworthi: Blank, vacant, inaneness perfect. Nulliebiety remarkable, visible only from certain points of view. Shadowless. An airy species whose flight cannot be followed except by the winged mind. From a naked chasm on Jupiter, August 5th, 1919."

He went on in another paper to describe a new species of human, Homo perniciosus, known only from the female sex, but by this time he had long since been himself classified as insane. I am not so sure. As one who has spent much of his life grumbling at editors, and as one who is aghast at the way contemporary "space biologists" muddy the line between fact and fiction, I cannot help wondering if Girault was really so mad as they claimed. Perhaps he should be regarded as the father of exobiology and the patron saint of the Society for the Extinction of Editors, which I intend to found one of these days.

But back to the bedbug. This bug belongs to a group that primarily attacks bats, and it is believed that it switched over from bats to men when both shared certain caves a few tens of thousands of years ago. Even today bugs often occur in churches that have bats in their belfries, taking advantage of whichever host is most readily available. Presumably they thrive best in churches having large belfries or long-winded preachers.

That bedbugs were well known in ancient Greece is shown by the fact that Aristophanes referred to them in at least two of his comedies. In *The Frogs,* Bacchus, about to descend into Hades to seek the release of Euripedes, asks for information from Hercules, who has just been there on business of his own. "Tell me about it," asks Bacchus, "the roads, the bridges, the brothels . . . the inns and taverns, and lodgings free from bugs and fleas, if possible." And in *The Clouds,* when Socrates asks Strepsiades to bring him a couch, the latter replies: "But I can't. The bedbugs won't let me."

Bedbugs have always been much at home in the Mediterranean area, and it is believed that they may have been carried to England by Julius Caesar and his legions. In England they were first called by their Latin name, Cimices, later corrupted to "chinches," and somewhat later by their German name, "wall lice" (*Wandläusen*). The word "bug" was not transferred to these insects until the seventeenth century, apparently from the Welsh word for "hobgoblin," which has also come down to us in such words as "bogeyman" and "bugbear." It is interesting that Shakespeare used the word in its original sense, as when Hamlet remarks "With ho! such bugs and goblins in my life." The first use of the word "bug" to apply to insects that haunt one's bed at night was probably a poet's metaphor, but by 1730, when John Southall published his classic *Treatise of Buggs* in London, the use was well established. Southall's interest, by the way, was not merely academic; on the title page he described himself as "maker of the Nonpareil Liquor for destroying Buggs and Nits." Samuel Pepys, incidentally, still referred to them as wall lice. In his Diary for June 12, 1668, he reported that in a small inn on the way to Bath he and his wife woke up to find their "beds good, but lousy; which made us merry." Pepys was always one to enjoy the finer things of life.

During the eighteenth and nineteenth centuries, many "destroyers of vermin" flourished in England, perhaps the most famous of whom were Tiffin and Son, "Bug Destroyers to Her Majesty." The Tiffins worked "for the upper classes only," and their clientele included dukes and princesses. The Tiffins became very knowledgeable in the ways of bugs, as revealed by these remarks of the elder Tiffin to Henry Mayhew in the 1860's:

"The bite of the bug is very curious. They bite all persons the same, but the difference of effect lies in the constitutions of the parties. . . . When nobody has slept in a bed for some time, the bugs become quite flat; and, on the contrary, when the bed is always occupied, they are round as a lady-bird. . . . Some people fancy, and it is historically recorded, that the bug smells because it has no vent; but this is fabulous, for they *have* a vent. It is not the human blood neither that makes them smell, because a young bug who has never touched a drop will smell.

"I know a case of a bug who used to come every night about thirty or forty feet—it was an immense large room—from the corner of the room to visit an old lady. . . . It took me a long time to catch him. . . . The finest and fattest bugs I ever saw were those I found in a black man's bed. He was the favorite servant of an Indian general. . . . His bed was full of them, no bee-hive was ever fuller."

England, of course, had no monopoly on bedbugs. In France, the bug was called *la punaise* (which might be translated "stinker"), in Egyptian villages *akalan* ("an itching"), in Sanskrit *uddamsa* ("biter"), and in Swahili *kunguni* (which is best not translated). However, none of the New World civilizations had a word for the bedbug, which suggests that the bug came to America from Europe; there are records that suggest that wooden sailing vessels of the sixteenth through the eighteenth centuries were almost always heavily infested with them. Peter Kalm, writing in 1749, remarked that the bedbug was common in the English colonies but unknown to the Indians. The fact that the bug's immediate relatives are inhabitants of the Old World confirms this origin for our good friend.

This information and much, much more is to be found in a recently published monograph on the bedbug and his relatives, authored by Robert L. Usinger, professor of entomology at the University of California at Berkeley (with chapters by various other specialists). There is doubtless many a bedbug that has devoted its life to the charms of an individual human who happened to provide a dependable blood meal when needed; but there are not many persons who have devoted the better part of their lives to the charms of the bedbug. For there is much to be said for the bug (personally I found some sections of Usinger's 585-page book almost too brief). Usinger has collected bugs all over the world and has reared many of them in his laboratories at Berkeley. Fortunately, most of them do well on rabbits, without bothering the rabbits noticeably, and since they require only one meal between each molt, their cage can simply be strapped to a partially shaved rabbit periodically. Since Usinger kept detailed pedigrees on most of the stocks he reared, things became a bit difficult when he traveled to scientific meetings or elsewhere. He solved this by simply carrying the necessary stocks of bugs in vials in his suitcase and strapping them on to his own arm when they required a meal. He does not say very much about this in his book, doubtless because he does not wish to become a persona non grata to hotel managers. In fact they need not worry; Usinger is as protective of his pedigreed bugs as a maiden lady of her poodle. Personally, I have always thought the poodle rather a travesty of a mammal. The bedbug is every inch a bug, and he does not yap or sniff at people's legs. (Well, at least he does not yap.)

He does, of course, stink a bit, and all the care in the world can do nothing to remedy that. The odor is so distinctive that an experienced person can often detect a heavily infested bedroom without seeing a bug. One person has described it as an "obnoxious sweetness." Most books describe it as a "buggy" odor, which seems a safe enough description, if not very informative. Tiffin and Son would be interested to learn that the odorous substances are discharged from special glands on the thorax. These glands resemble in a general way those of typical "stinkbugs," most of which are plant feeders, and the chemistry of the odorous materials is similar, though not identical. We know that the scent glands of stinkbugs protect them against birds and other predators. Of what possible use can they be to the bedbug, which is secretive in its habits and almost never encountered by birds? Perhaps they protected them from being eaten by their original hosts, bats. Experiments have shown that bats will not eat bedbugs even when they are stuffed into their mouths. Mealworms, which bats normally eat with relish, are rejected when smeared with the scent of bedbugs. The bedbug no longer needs to stink, of course, since humans have little appetite for such things anyway. He just can't help himself. It is true that bedbugs were once recommended as a cure for certain types of infections. Mixed with salt and mother's milk, they made an ointment for the eyes, and seven bedbugs eaten with beans were supposed to relieve certain types of fever. But as food—never.

It is known that odor plays no role in the love life of bedbugs. Males apparently find the females merely by blundering upon them, and they will attempt to mate with a piece of cork carved in the shape of a bug. It is also known that they find their source of blood by a simple wandering behavior combined with an attraction to warmth. It is as though they could not smell at all, doubtless an invaluable adaptation in an animal that is so at home in brothels and almshouses and smells so badly himself.

There is, I am sure, no connection between the bedbug's love of

The tail end of a male bedbug, showing the large, hooklike penis, which is retracted into a groove when not in use. With this scimitar the male punctures a hole in the body wall of the female at a point far from her true genital opening.

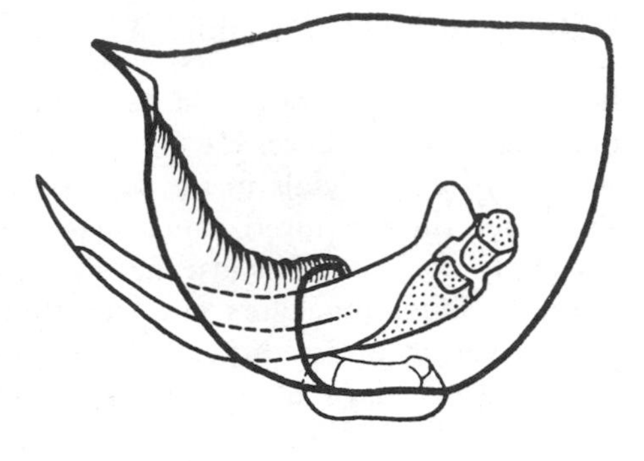

brothels and his own fantastic sexual deviations. Mating in this insect has been termed "traumatic insemination," meaning simply that the male punctures a hole in the female and inserts his semen there. His intromittent organ is a stout hook, suggesting half of an ice tongs. The hook is inserted through the membrane between the fifth and sixth abdominal segments of the female. Here there is a small notch that guides the male penis into a swollen mass of tissue called the "organ of Berlese," after the Italian entomologist who discovered this structure and who has also had a collecting apparatus named after him, as mentioned in Chapter 2. There is, however, no opening to the outside, and the male actually punctures the body wall of the female. Following mating, the wound heals over and a scar forms. The female can keep no personal secrets from the entomologist, who has only to count the scars to find out how often she has mated.

The male bug injects an unusually large amount of semen into the female. The organ of Berlese apparently acts as something of a pad to protect the internal organs from laceration. It also helps to prevent bleeding, assists the healing of the wound, and absorbs much of the seminal fluid. The sperms themselves migrate to the wall of the organ of Berlese and then move into the blood of the bug. Like most animal sperms, they swim actively. It appears that insect sperms swim at roughly the same speed as human sperms (nearly an eighth of an inch per minute, which is pretty fast for something invisible to the naked eye). The sperms make their way through the blood to the true reproductive organs of the female, finally gathering in little sacs called sperm reservoirs. Here they remain until the female takes a meal, whereupon they migrate up the egg tubes to the ovaries, ready to fertilize the eggs that are rapidly fabricated from the nutrients in the blood of the host.

This unique method of mating occurs in quite a number of parasitic and predatory bugs related to the bedbug, though hardly any two species have the organ of Berlese in exactly the same place (much to the liking of persons involved in classifying these animals). It is said that in one species occurring in bat caves in Texas and Guatemala, the male jabs the female almost anywhere on the abdomen, there being no special slit or organ of Berlese. The sperms then migrate to the heart, where they are pumped into all parts of the body, even into the head and the ends of the legs. Finally they collect in sperm reservoirs as in the bedbug, then migrate to the ovaries.

A female bedbug, seen from beneath. The notch on one side of the fourth segment behind the legs serves to guide the penis of the male into the organ of Berlese. From this organ the sperms must swim through the blood to the female genital system.

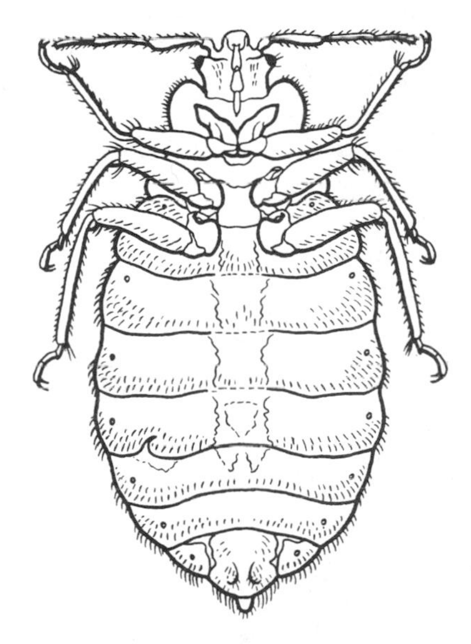

There have been a good many theories as to the value of this weird system. Dr. Berlese was impressed by the great number of sperms introduced by the male and by the fact that at least some of the semen appears to be absorbed by the female. He suggested that the female may actually supplement her diet in this way, perhaps even to the point of being able to lay eggs in the absence of a blood meal. Quite recently Jacques Carayon, of the Natural History Museum of Paris, a man who has spent many years studying the sexual perversions of these insects, found that the males of an African species habitually inseminate each other. This seems to support the idea that bugs are able to transfer protein-rich nutrients to one another, possibly an important adaptation in an animal that sometimes has to wait long periods for a blood meal to become available. The image of a covey of bedbugs disporting themselves in this manner while waiting for a blood meal—copulating with either sex and at the same time nourishing one another with their semen—makes Sodom seem as pure as the Vatican. But it is probable that we are maligning the bedbug. The absorption of unused sexual products occurs in all animals, and the bedbug may simply have to inject more semen than usual because of the loss of sperm in this devious system. And we all know that homosexuality is not restricted to the bedbug. We must set aside Berlese's interesting theory as unproved, but having done so we are left with no explanation as to the advantages of traumatic insemination. Needless to say, it is not allowable to credit a mere bug with inventing all this simply for the sheer ecstasy of it.

The bedbug has still other unusual adaptations one would not expect in so disreputable a beast. It happens that blood—the only food it ever takes—is a wonderful diet except that it lacks B-group vitamins, which are required by insects. All bloodsucking bugs have apparently solved this problem by working out a partnership with certain bacteria that produce B vitamins in quantities. If one dissects a bedbug, he finds a pair of little ovoid white bodies in the abdomen. These contain bacteria that are capable of living nowhere else and that "pay for their keep" by producing vitamins needed by the bug. These bacteria are passed on to the next generation via the egg, each one of which apparently contains enough to start a colony in the offspring.

When feeding, the bedbug, like other bloodsuckers, produces a substance from its salivary glands that prevents the blood from clotting for as much as twenty-four hours—by which time digestion is well advanced. It is apparently this salivary fluid that produces a reaction in some persons. As Mr. Tiffin of nineteenth-century London was well aware, the bugs "bite all persons the same," but people vary greatly in their reactions. The bite itself is painless, an important requisite for an insect that feeds on sleeping persons. Some persons show no aftereffects at all, and others develop immunity after being bitten repeatedly. Professor Albrecht Hase, of the University of Jena, Germany, permitted himself to be bitten by 2,500 bugs over a nine-month period, and thereby acquired immunity to their bites. On the other hand, Robert Usinger fed a colony on his own blood every week for seven years, and still developed itchy welts at the points of feeding. Like a true taxonomist, Usinger allowed various bat bugs and bird bugs to bite him in order to compare the effects of their saliva with that of the bedbug. He found that species closely related to the bedbug produced a reaction similar to that of the bedbug but that more distantly related species produced different kinds of reactions, more or less in accordance with their supposed relationships.

Books speak of the bite of the bedbug as "benign," meaning that it is painless and capable of producing no more than a mild reaction. Be this as it may, people have been devising ways to escape bedbugs for many years. Democritus is said to have recommended hanging the feet of a stag at the foot of the bed. A more useful remedy proved to be the placing of each leg of the bed in a bowl of water, but this simply prevented the bugs already living in the mattress or bedstead from going elsewhere. Furthermore, many reputable persons have reported that "the indomitable little devils will climb up to the ceiling, survey the spot, and drop vertically upon their victim below," to use the words of Malcom Burr, in *The Insect Legion.* Personally, I wonder if their powers of "surveying the spot," at least, are not exaggerated, but Burr claimed to have observed this behavior. He should know, for after spending several years in Russia he remarked, "I believe there is not a house in Russia, in the whole Soviet Union, which is free from them." This was nearly half a century ago, at a time when four million people in London were plagued with bedbugs and seven hundred exterminators were said to be active in Germany. By this time various fumigants and oil sprays had replaced the secret formulas of Southall, Tiffin, and other earlier specialists, but bedbugs had a way of promptly reinvading rooms cleaned out by these methods.

DDT changed all that, beginning about the time of World War II, for surfaces sprayed with that "wonder insecticide" remained toxic to bugs for several months. Unfortunately, the situation has deteriorated somewhat, as strains of bedbugs have evolved that are resistant to DDT and to several other insecticides. One strain has remained capable of living in the presence of DDT even after nearly five years of non-exposure to DDT or other insecticides (most insects lose their resistance after several generations of non-exposure). However, the situation is not desperate, as most bedbugs still succumb to good household sprays, and modern furnishings and cleaning practices are not always what the bedbug would prefer.

Robert Usinger remarks in his book that "an entirely different approach to bedbug control was suggested" by experiments of certain workers "who fed sublethal doses of DDT and pyrethrum to rabbits and found that bedbugs died after a few hours when fed on them." I am not sure whether the suggestion is that persons bothered with bedbugs consume sublethal doses of DDT or that they keep a pet rabbit in their bedroom and slip it a little DDT now and then.

In either instance, I would just as soon have the bedbugs. After all, they aren't so bad. Not only is their bite "benign," but after many years of trying, entomologists have been unable to incriminate the bedbug with the transmission of any human disease whatever. The flea transmits plague, the louse typhus, the mosquito malaria, the tsetse fly sleeping sickness. But the bedbug is as innocent as a lily, even though his odor may not suggest it.

Unfortunately, these remarks do not apply to all bloodsucking bugs. There is a large group, somewhat distantly related to the bedbug, generally called the "cone-nosed" bugs because the front of their head tapers off like a cone. They are also called "assassin bugs," since the majority kill and suck the blood of other insects. Many of the insect-feeders bite humans when they are handled or accidentally touched; their bite is far from "benign": it can be incredibly painful, and may cause swelling or pain that persists for weeks. I remember very well my own first experience with one of the assassin bugs. We were collecting the insects attracted to a Coleman lantern in a park in Texas when I brushed at something that had landed on the back of my neck. There was a sudden shot of pain that carried right down to my shoes, and my neck and shoulders became so sore that I was unable to swing an insect net properly for several days. In this case the bug was a species known as the "two-spotted corsair," though that is not the name I used for it that evening.

One of the assassin bugs often occurs in houses, and is said to feed on its cousin the bedbug. It is covered with a sticky substance, and as it crawls about it becomes covered with dust and lint. In this disguise it is called the "masked bedbug hunter." When roughly handled, it is capable of inflicting a severe bite. A number of people have been bitten around the lips, presumably when they brushed off the insect as it was hunting bedbugs at night. This has earned it the name "kissing bug."

According to Malcolm Burr, the emirs and khans of central Asia used to keep cone-nosed bugs to torture their prisoners. Two British diplomats were said to have been thrown into a bug pit by the Emir of Bokhara in 1842. Here they remained for several months before they were taken out and beheaded. The Emir is said to have thrown chunks of fresh meat into the pits for the bugs to feed on when no prisoners were available.

There are a number of cone-nosed bugs that have taken to feeding on warm-blooded animals regularly, and these have developed the ability to suck blood without disturbing the host. One such insect is called the Mexican bedbug, since it is especially prevalent in adobe huts and other dwellings in Mexico, although in fact it ranges widely in the southern United States also. This species has many relatives in Central and South America, all of them considerably larger than the common bedbug, and many of them addicted to feeding on the blood of man at least some of the time. Several of them are known to transmit a very serious although poorly understood disease of man, called Chagas' disease or American trypanosomiasis. In *The Voyage of the Beagle,* Charles Darwin told of his encounter with a member of this group in a village near Mendoza, Argentina:

"At night I experienced an attack (for it deserves no less a name) of the Benchuca . . . the giant black bug of the Pampas. It is most disgusting to feel soft wingless insects about an inch long, crawling over one's body. Before sucking they are quite thin, but afterward they become round and bloated with blood, and in this state are easily crushed. One which I caught . . . was very empty. When placed on a table, and though surrounded by people, if a finger was presented, the bold insect would immediately protrude its sucker, make a charge, and if allowed, draw blood. No pain was caused by the wound. It was curious to watch its body during the act of sucking, as in less than ten minutes it changed from being flat as a wafer to a globular form."

Darwin found these bugs to be very common in parts of South America. Incidentally, one theory regarding Darwin's persistent illness during much of his life is that he picked up Chagas' disease unknowingly on this trip. One form of the chronic condition of the disease is characterized by nervous symptoms similar to those of Darwin in his later years. However, there are several other theories regarding the origin of Darwin's illness, and it is far from certain that Chagas' disease was involved.

The causative organism of Chagas' disease is a minute one-celled animal (trypanosome) that swims actively with a whiplike tail; it is related to the causative agent of African sleeping sickness. This organism goes through part of its life cycle in the intestine of the bug. When the bug feeds on a warm-blooded animal, it has the habit of defecating at the same time. When the bite is later scratched, some of the infected feces may be rubbed into the wound or carried to the eyes or mouth. Entry through the eyes produces a swelling of the eyelids and face, and entry at any point produces persistent fever and anemia. The mortality rate is only about 5 per cent, but patients who survive normally develop a chronic heart or nervous condition because of the prolonged survival of the trypanosomes in their system. Unfortunately, the symptoms resemble those of several other diseases, and the infective organisms are difficult to find in the bloodstream. One method of diagnosis is to take a noninfected cone-nosed bug and allow it to engorge on the patient; if the latter has the disease, the trypanosomes will build up in the intestine of the bug and can easily be found there or in its feces.

Chagas' disease occurs throughout most of South and Central America, and examination of bugs from various places has shown from 20 to 50 per cent to be infected with trypanosomes. One population of bugs in Texas was found to be 92 per cent infected, and a typical case of the disease was produced experimentally from Texas bugs. However, very few authentic cases of Chagas' disease have been reported from the United States, probably because of our relatively sanitary living conditions and, perhaps, the unfamiliarity of physicians with this exotic disease and their consequent misdiagnosis of the symptoms. Many animals serve as reservoirs of the infection, especially armadillos, opossums, bats, rats, mice, and even cats and dogs. Probably many human infections come from bugs that were themselves infected by feeding on one of these animals. It seems to be characteristic of cone-nosed bugs that they are not at all fussy about what animal they feed upon. Blood is blood to them.

Like the true bedbug, the cone-nosed bugs have bacteria in their bodies that apparently produce vitamins they do not obtain in their blood diet. In contrast to the bedbug, the cone-nosed bugs have a fairly normal and unimaginative mating behavior. They do have it over the bedbug in two respects. For one, the adults can fly, and thus disperse themselves widely. For another, the beak of these bugs is stout and rigid, and its tip can be rubbed over a series of parallel ridges on the underside of the thorax to produce a squeaking sound. Since this sound can be produced by both young and adults, and since it has an irregular pattern and is not notably different in the various species, we are pretty sure that it is not a mating call similar to that of the cricket. The squeaking is most often produced when a cone-nosed bug is disturbed or seized, and we assume from this that it is a warning sound, that is, that it serves to startle birds or other predators and cause them to drop their intended prey.

The voice of the bug has recently received a considerably amount of publicity. A large Oriental species has been adopted by the United States Army, dubbed the "combat bedbug," and used experimentally as a device for tracking down enemy troops in the jungles. According to newspaper reports released in June, 1966, the bug is carried in a capsule that "allows him to smell out a man about two blocks to the front or sides but not the trooper carrying him." When the bug senses "the nearness of human flesh," he lets out a "yowl," which, however, has to be amplified to be audible to the human ear. The bug is said to have been field-tested under "Vietnamlike" conditions. Presumably it will be carried by advanced patrols and will warn the troops about Vietcong lying in ambush. All this sounds a bit fantastic, and I can only say that, like Will Rogers, "all I know is what I read in the papers." It seems ironic that a decade after the development of the hydrogen bomb we are calling on bugs to help us fight our wars.

The cone-nosed bugs may not revolutionize the art of warfare, but they have played an exceedingly important role in helping us to understand animal growth and metamorphosis. This is because they are easy to rear and feed in the laboratory and because a culture of them, some forty years ago, came into the hands of a biologist shrewd enough to appreciate their usefulness in solving some of the critical problems of insect development. The species of bug was Rhodnius prolixus, an inhabitant of the American tropics that is known to carry Chagas' disease. About fifty years ago a culture of this bug was acquired by the London School of Hygiene and Tropical Medicine, where it came to the attention of a young research worker named V. B. Wigglesworth. Beginning in 1933, Wigglesworth published a series of papers on the physiology of Rhodnius that have become classics in their field. Now Sir Vincent Wigglesworth, he has recently retired as Quick Professor of Biology at the University of Cambridge, and is the author of innumerable research papers and several excellent books. It would seen unflattering to say that Sir Vincent has ridden to fame on the back of a bug, so I shall say instead that he has made one of the most disreputable of creatures the measure of all things insectan. It sometimes seems that every other paragraph in entomology begins: "Now as Wigglesworth has found in Rhodnius . . ."

There is really nothing special about Rhodnius; he simply happened to be at the right place at the right time. The adult is about three fourths of an inch long and of rather drab coloration. Like the bedbug, Rhodnius molts five times during its growth from hatchling to adult, requiring one large blood meal between each molt. Feeding techniques in the laboratory were described quaintly by P. A. Buxton, of the London School of Hygiene and Tropical Medicine, in 1930.

"It is convenient to use a lop-eared rabbit; if only a few bugs are to be fed . . . they can be put in a short wide tube, the mouth of which is covered with gauze, and attached to the ear of the rabbit by an elastic garter. If many bugs are to be fed, or if they are large, they may be put in a jampot, into which the ear of the rabbit is introduced; the mouth of the pot and the folds of the ear are then packed with cotton wool. The rabbit will sit quietly at the bottom of a box, 7 inches wide by 15 inches long; this gives it room enough for meditation, but none for exercise. It appears that the rabbit does not feel the punctures, and it gets no subsequent reaction. We have a rabbit on which dozens of bugs have been fed every week for more than a year, and on which we have sometimes fed more than a hundred in a day; but its ears are neither red nor thickened."

When the hatchling Rhodnius feeds, it takes in more than ten times its own weight in blood, its relatively soft, distensible abdomen becoming enormously swollen. The larva, after the first molt, takes in five to seven times its own weight of blood, but the adult (after the fifth molt) has a less distensible body, and is able to take only one and a half to three times its own weight in blood. The adults normally feed several times, and the females lay several batches of eggs. It is interesting that the great weight increase of the engorged bug lasts only a short while; within three or four hours most of the water in the blood, along with some other unwanted constituents, is excreted by the bug, bringing its weight down to only slightly more than it was before the meal (or no more at all, in the case of the adult). Complete digestion of the blood takes several weeks, and in the meantime an immature bug has undergone a molt that grants it a slightly larger body size and the capacity to take a slightly larger blood meal the next time. We may feel fortunate that something "shuts off" growth and molting after the fifth molt; and we may feel curious, as Professor Wigglesworth did, as to what triggers a molt after each larval blood meal but fails to trigger a molt after an adult engorgement. The close correlation between feeding and molting in Rhodnius made this insect a much more appropriate animal in which to study these things than one that feeds continuously, for example a caterpillar.

As a matter of fact, when Wigglesworth started his work some earlier studies on caterpillars were already strongly suggestive. Stefan Kopeć, of the Agricultural Research Institute of Pulawy, Poland, had developed a technique of removing the brain of gipsy moth caterpillars without causing their death. In 1917, he demonstrated that caterpillars made brainless soon after molting failed to molt again, although they lived well beyond the time when they normally would have molted. At the same time, caterpillars with other parts of the nervous system removed did molt normally, as did those with brain operations in which the brain was left in the body. He suggested that the brain produced a chemical that circulated in the blood and induced molting.

Wigglesworth discovered that Rhodnius bugs decapitated within one day after feeding would (if the wound were sealed with wax) live for long periods, occasionally even for a year—much longer than they would have lived if they had "kept their heads." However, they never molted. This might, of course, have been the result of the fact that they did not feed and thus did not undergo the stretching of the body wall that might trigger molting. However, when Wigglesworth transplanted brains into the abdomens of these decapitated individuals, they molted. This confirmed Kopeć's belief in a "brain hormone."

Both Kopeć and Wigglesworth found that if the brain were removed after a certain time had elapsed—in the case of Rhodnius, three to eight days after feeding—"debraining" no longer prevented molting. Apparently by this time the brain has already released its hormone and is no longer essential to molting. To demonstrate that this is so, Wigglesworth cleverly joined two decapitated bugs, neck to neck; one was decapitated one day after feeding, the other eight days after feeding. When the latter molted, the former did also, even though it would not normally have done so. However, the brain hormone, already present in the second individual, circulated into the first individual and caused it to molt.

It was soon discovered that the whole brain was not involved in the production of this hormone, but only certain large, glandular cells embedded in it. However, conflicting evidence regarding the control of molting was accumulating. Certain glands had been described in the thorax of caterpillars in the classic work of Pierre Lyonet in Holland in 1762, but little attention had been paid to these glands until the 1940's, when the Japanese biologist Soichi Fukuda showed that silkworms will molt into the pupal stage only when these thoracic glands are present. Could this information be reconciled with the evidence that it was cells in the brain that produced the hormone that brought about molting? Carroll M. Williams, of Harvard University, working with the cecropia moth, performed a brilliant series of experiments, beginning in 1942, which proved that both brain hormone and thoracic gland hormone are essential to molting. Professor Williams implanted brains into abdomens that had been cut off from pupae from cocoons of the cecropia moth, but obtained negative results. However, when thoracic glands were also implanted, the abdomens molted into the adult stage. If the abdomen was that of a female, it would even attract and mate with a male and produce eggs! Further experiments showed conclusively that the brain hormone acts upon the thoracic glands, stimulating them to produce the molting hormone.

Thoracic glands were soon discovered in Rhodnius, and it was found that the whole process is very much the same as it is in moths. In his early experiments, Wigglesworth had left the thorax intact, so that the thoracic glands were functional. He now tried implanting brain substance into isolated abdomens, and found that they would not molt; but if he implanted thoracic glands that had been activated by brain hormone into isolated abdomens, molting ensued. In Rhodnius, it is the stretching of the body wall after feeding that "informs" the brain, via sensory cells and the nervous system, that it is time to molt. Wigglesworth showed that bugs in which the abdomen was artificially stretched were thereby induced to molt; but when the nerve cord was severed, the message did not reach the brain, and no molting occurred. Ordinarily, when the message reaches the brain, its hormone is released, and this in turn causes the thoracic glands to become active and to produce a substance that travels via the blood to the cells of the body wall and sets in motion a complex series of processes that result in the old body wall being sloughed off and a new one laid down. Why does molting cease when the insect reaches the adult, sexually mature stage? Simply be-

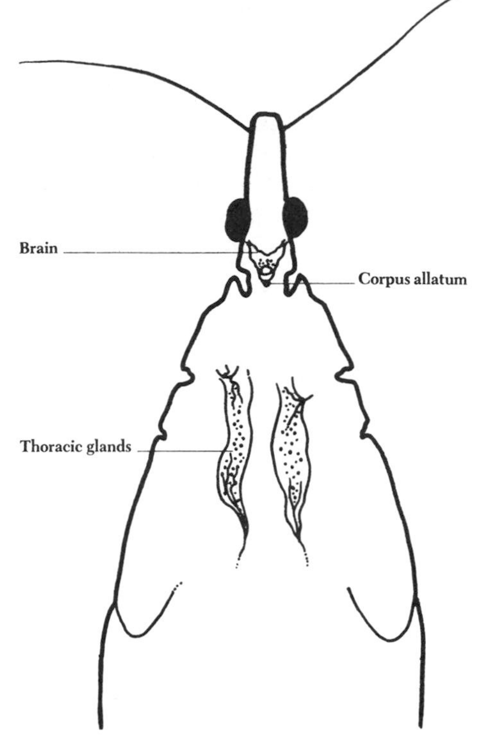

The endocrine organs of the cone-nosed bug Rhodnius prolixus. (After Sir Vincent Wigglesworth.)

cause adulthood coincides with the degeneration of the thoracic glands. This being true, it should be possible to implant active thoracic glands into an adult and cause it to molt again. Wigglesworth showed that this can, in fact, be done.

Biologists soon became curious as to the nature of this hormone, but it occurred in such minute amounts that it seemed impossible ever to obtain enough so that the substance could be identified chemically. Then Peter Karlson and his colleagues at the Max Planck Institute in Munich, Germany, decided to undertake a "crash program" to identify the hormone. They purchased live silkworms from all over the world, and eventually obtained about four tons of them. From this great mass of caterpillars they obtained 250 milligrams (not quite 1/100 of an ounce) of crystalline molting hormone. They analyzed it chemically and found it to be related to cholesterol, a substance occurring widely in animal tissues and one that has recently received much publicity, since its accumulation in human blood vessels is believed to be a cause of circulatory failure. Karlson proposed the name "ecdysone" for this substance, although most of us are happy enough to go on calling it the "molting hormone," conveniently abbreviated MH. This hormone is the same in all insects; silkworm MH will work not only on the silkworm but also in flies, Rhodnius, and other insects.

The step-by-step unraveling of some of the mysteries of insect growth has been one of the great but little-known dramas of recent years. But of course knowledge of the molting hormone does not answer all the questions that come to mind. What causes Rhodnius, at the last molt, to turn into a rather different-looking insect? The males and females, which were formerly indistinguishable, are suddenly found to have very different genital organs; ocelli (simple eyes) suddenly appear on top of the head; the legs develop adhesive organs; the thorax is differently shaped and bears fully developed, functional wings; and the abdomen is differently shaped, differently colored, and its surface loses the "pockmarked" appearance caused by the presence of hundreds of small sense organs in the larva. We think of the adult as being *the* insect; but why and how have the adult structures remained suppressed throughout the long larval period? If we can explain this, it should be possible to go further and explain the still more dramatic change that occurs when the maggot becomes a fly, the caterpillar a butterfly. In the words of Sir Thomas Browne, it is as if there were "two Souls in those little Bodies." Wigglesworth puts it this way:

"The contemplation of the metamorphosis of insects has always evoked feelings of mystery. When regarded more closely through the eyes of the anatomist and the experimental biologist, the superficial mystery is dispelled—to be replaced by deeper mysteries."

Insects are truly dual personalities; they are born with the capacity to be two insects, one following the other. Both personalities are inherited from their parents, but the definitive one—that of the adult, reproducing animal—is prevented from expressing itself until the final molt. The advantage of this is obvious: the larva can become a specialist in eating, the adult in reproducing. There is no more effective eating machine than the caterpillar devouring a leaf; and no more effective reproducing machine than the female moth, who attracts her mate from great distances with her pheromones prior to laying great numbers of eggs on the proper food plant. In the case of the bloodsucking bugs the larva is less spectacularly different from the adult, but as we have seen he is much more "stretchable," and capable of consuming much more blood in relation to body weight than is the adult; and the blood he digests goes into building up his own body size and food reserves rather than being deployed in the production of sperms and eggs. All the adult features are latent in the larva; indeed, each cell carries the capacity to assume the form characteristic of the adult. What controls this suppression and sudden expression of adult features?

When Wigglesworth cut off the head of Rhodnius larvae soon after feeding, they failed to molt, as we have seen, and when he cut off the head of larvae several days after feeding, they molted, since the thoracic glands had already been activated by the brain hormone. However, he found that there was a critical period—three or four days after feeding—when small larvae that were decapitated often molted not into the next larval stage but into diminutive adults! Clearly, whatever the factor is that prevents the expression of adult features, it is associated with the head; and since it affects all the body cells it is certainly a hormone carried by the blood. Wigglesworth determined that this hormone emanated from a small mass of cells just behind the brain, the corpus allatum. When he removed the corpus allatum from a fourth-stage larva and put it into the abdomen of a fifth-stage larva, the latter molted not into an adult but into a giant sixth-stage larva. Some of his sixth-stage larvae molted still another time and produced giant "superbugs." Quite obviously this hormone—the juvenile hormone (JH) as it came to be called, since it maintained juvenile features—dropped out of the picture just before the final molt, permitting sudden expression of adult features. This nonfunctioning of the corpus allatum after the next-to-the-last molt is evidently programmed by heredity, for the number of molts an insect undergoes is often quite fixed. Even Professor Wigglesworth was not able to alter the number of molts and time of metamorphosis indefinitely, though he did provide the foundation for a new genre of science fiction: a world filled with gigantic insects. We may hope it remains in the realm of fiction, and there is every reason to believe it will, for insects simply do not have mechanisms for breathing or body support that will permit overly large size.

It is impossible to do full justice here to all the exciting experiments that have been performed in the course of clarifying the function of insect hormones. I shall only mention one or two of Wigglesworth's more unusual ones. When the source of the juvenile hormone had definitely been pinned down, it was decided to see if its effects could be produced externally, that is, if one could apply the extract of the corpus allatum to the outside of the fifth-stage larva so that in the ensuing molt an adult with larval patches would be produced. Wigglesworth found that this was possible if he abraded the integument so that the substance would penetrate to the cells of the body wall. In this way he produced an adult with one normal wing and one larval wingpad. In another experiment he etched his initials with corpus allatum extract on the back of a fifth-stage larva. The resulting adult was normal except for the initials VBW spelled out beautifully in dark, speckled larval integument against the pale, smooth integument of the adult!

Multiple joining of Rhodnius larvae and adults also produced interesting results. When two fifth-stage larvae were joined to an adult, the adult was caused to molt to a second adult stage. But when two more larvae, this time of the fourth stage, were also joined to the adult, the adult molted not to another adult but to a polyglot of adult and larval

Multiple joining of Rhodnius. Two fourth stage larvae (*above*) and two fifth stage larvae (*left and below*) have had their heads cut off and have been joined to an adult, causing it to molt into a polyglot of adult and larval features. The junctions are sealed with paraffin (*black*). (After Sir Vincent Wigglesworth.)

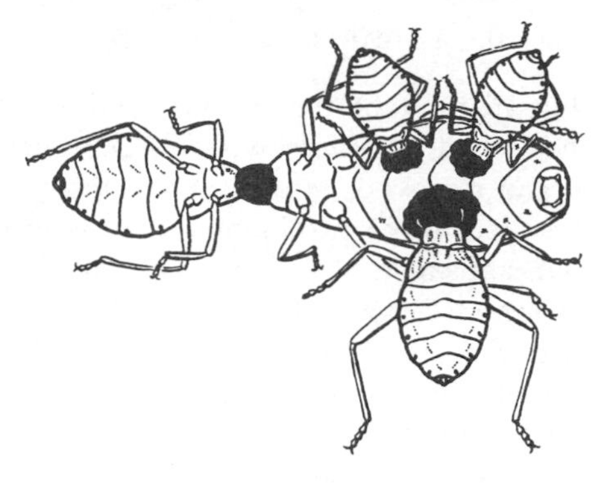

features. The fifth-stage larvae had no juvenile hormone in their blood, of course, but the fourth-stage larvae did. At this point one should pause and reflect on what wonderful creatures insects are for experiments of this kind. Imagine trying to unite five dogs in this way!

In the meantime, other workers were studying the juvenile hormone in insects exhibiting complete metamorphosis. Soichi Fukuda, in Japan, successfully removed the corpora allata from small silkworms and found that they spun tiny cocoons from which tiny moths developed. Fukuda and others showed that for the molt from larva to pupa a small amount of juvenile hormone must, however, be present, although the final molt to the adult is made in the absence of juvenile hormone. Carroll Williams, at Harvard, implanted active corpora allata into cecropia moth pupae, and caused them to molt to a second pupa. Thus it appears that "complete metamorphosis," as it occurs in moths and many other insects, is not different in any important way from the "incomplete metamorphosis" of bugs and the like. The caterpillar has evolved so far in a direction different from that of the adult that the stage between the last two molts has become a resting stage, the pupa, permitting large-scale reorganization to the adult form. The molt to the pupa is controlled by a certain decreased titer of juvenile hormone, while that to the adult is undergone in the absence of this hormone.

Curiously, the corpus allatum is known to resume its production of hormone a short time after the insect resumes its adult form. One can take the gland from an adult Rhodnius and implant it in a fifth-stage larva, causing the latter to molt into a sixth-stage larva, just as if one had used the gland of a fourth-stage larva. But in the adult the juvenile hormone does not cause a reversion to juvenile features: it conveys quite different messages in the body. Wigglesworth found that decapitated adult female Rhodnius never developed mature eggs in their ovaries. However, if he were careful to cut off the head behind the brain but in front of the corpus allatum, most of them did develop eggs. If he joined a female with the corpus allatum intact with one with this organ removed, circulation of the blood between the two in most cases resulted in the production of eggs in both. In the male, the hormone did not appear to be necessary for the formation of sperms, but it was required for full development of organs that contribute to the semen. We now know that this hormone influences other events in the life of the insect—for example, the production of sex pheromones, as we saw in the cockroach in Chapter 3; and the control of the "phases" of locusts, as we shall see in Chapter 10.

The growth and metamorphosis of Rhodnius. Each larval stage feeds only once. The distension of the body wall sends a stimulus to the brain, causing the release of brain hormone (*bh*). This stimulates the thoracic glands to release molting hormone (*mh*). This hormone produces a loosening of the body wall, which soon splits and permits the insect to crawl out. The new body wall is at first elastic, and the insect takes in air and swells to a slightly larger size; soon it will be ready for another and larger blood meal, followed by another molt. The thoracic glands

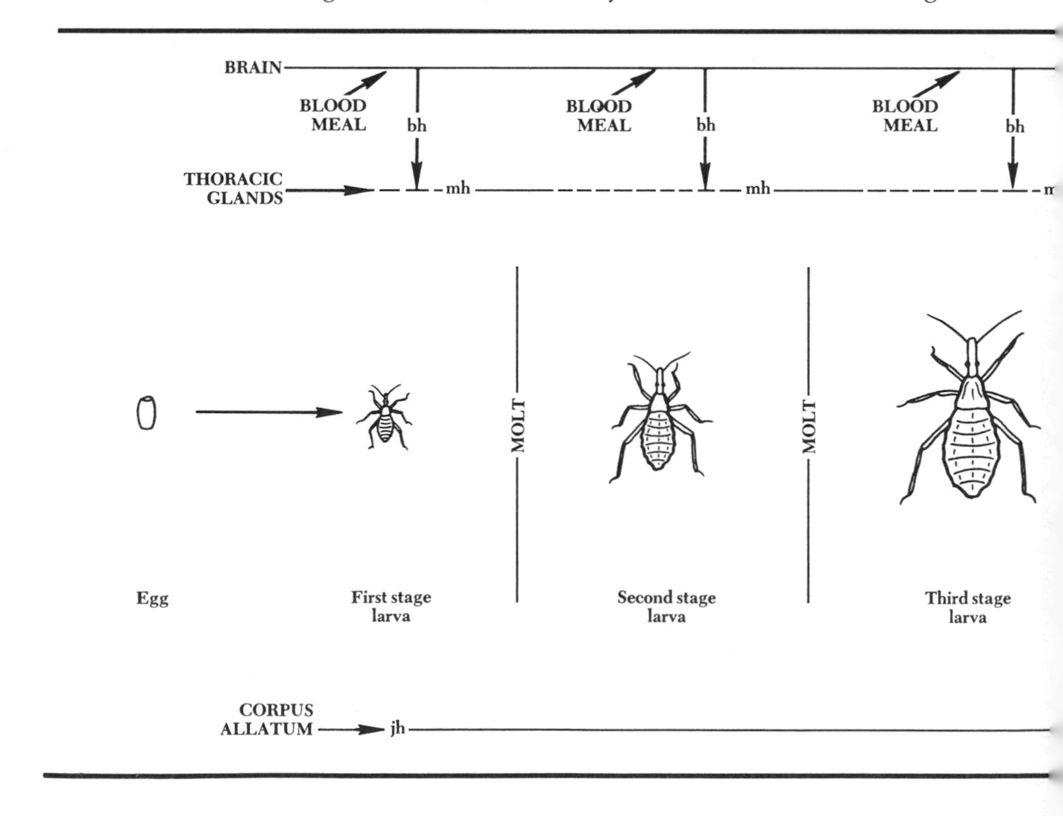

Since juvenile hormone is present and active in very minute amounts, like the molting hormone, there have been problems in obtaining it in amounts sufficient for experiments and for identification. These problems were in large part solved by Carroll Williams, who found that adult cecropia moths tend to accumulate the substance in considerable amounts in their abdomens. Experiments showed that it was derived from the corpus allatum, and simply stored in the fat by the males (but not the females). Since the cecropia moth is a large insect obtainable in fair quantities, a rich source of juvenile hormone was at last available. Williams' initial extract was a brownish oil of considerable potency,

no longer function after the adult stage has been reached. The corpus allatum produces juvenile hormone (*jh*) through the fifth larval stage, then becomes inactive, permitting a molt to the adult stage; somewhat later it resumes activity, and controls the development of the eggs of the female. Further blood meals provide the substance for the formation of the eggs, but there are no further molts, since the thoracic glands are non-functional.

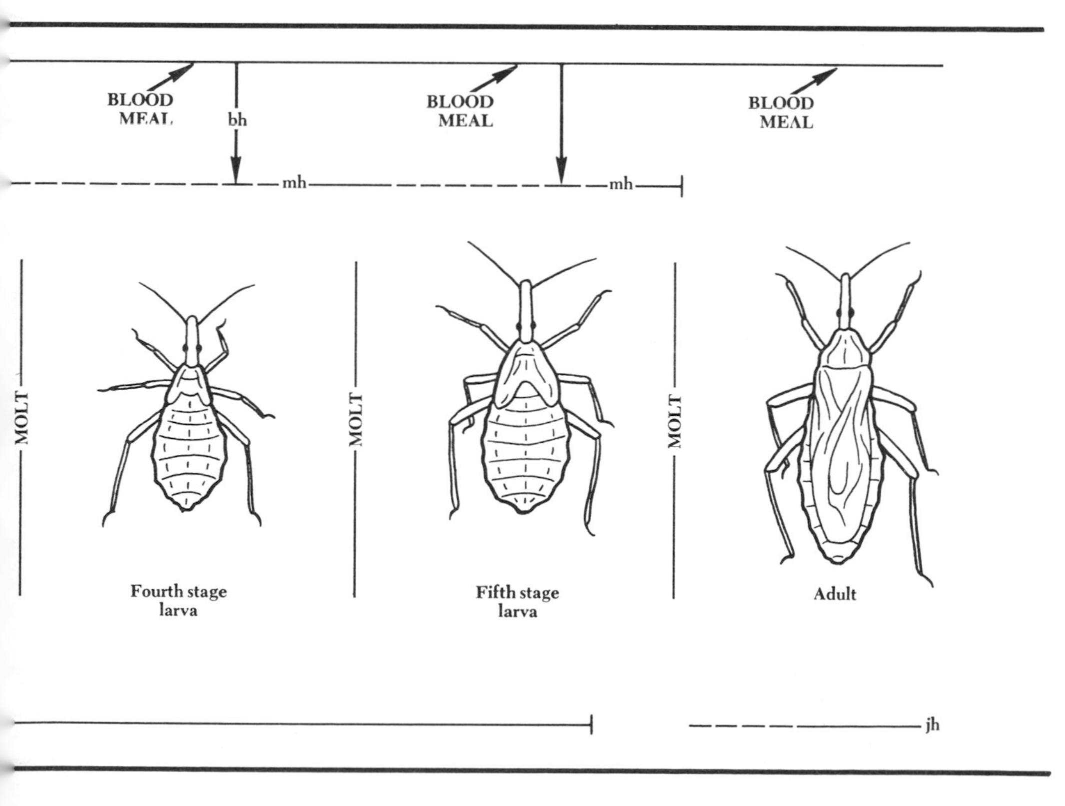

though by no means the pure hormone. In the attempt to obtain a more purified substance, he reared large numbers of cecropia caterpillars on wild cherry trees covered with netting, and also purchased cecropia cocoons from all possible sources. Over a period of weeks, he accumulated 1,000 adult male abdomens in a flask of ether in his laboratory. The abdomens were then placed in fresh ether and homogenized in a Waring blender. From this ethereal brew he eventually obtained a clear, golden oil that was then subjected to a series of sophisticated chemical treatments resulting in purification "about 50,000-fold over the original highly active extract of male Cecropia abdomens." Very recently Herbert Röller, of the University of Wisconsin, has succeeded in isolating and identifying the hormone itself, and as of this writing he has just announced its synthesis.

There are two fascinating "sideshows" to this drama, both involving substances derived from other sources that "mimic" the effect of the juvenile hormone. A few years ago it was discovered that a substance with "juvenile hormone effects" occurs widely in different animals, including vertebrates, and even in yeasts, bacteria, and soy beans. This substance is farnesol, a chemical regarded as of little importance in itself except as a halfway stage in the natural synthesis of certain fatty substances. Wigglesworth, in his book *The Life of Insects*, remarks that the insects appear to have "captured" a common substance of little biological significance and used it as a chemical messenger in the body, evolving a special organ, the corpus allatum, to produce it. Since the molting hormone is related to cholesterol, perhaps this is another example of "hormone capture" of a common biological substance. There may be some truth in this idea, but of course MH is not actually cholesterol, and JH is in fact far more potent than farnesol. Recently Williams and his colleagues have developed a substance with powerful juvenile hormone effects by bubbling hydrogen chloride gas through a cold solution of farnesenic acid. This substance could probably be produced commercially with ease and would prove a potent insecticide. One part of the substance in 100,000 parts of water was shown to block the emergence of yellow-fever mosquitoes, and small quantities applied to woolen pads prevented body lice from reaching sexual maturity, although some of them continued to molt into giant but impotent larvae ("loused-up lice," as *Scientific American* headed a news item). (The body louse, which transmits typhus and several other human diseases, has developed resistance to DDT and related insecticides.) This synthetic JH-analogue also prevents lice, caterpillars, and other insects from emerging from their eggs properly. The true juvenile hormone is still more powerful; Herbert Röller and his associates report that one gram (1/28 of an ounce) should prove sufficient to kill about a billion mealworms. It is difficult to imagine insects developing resistance to their own hormones; and it is probable that these substances will prove completely nontoxic to vertebrate animals. Of course, these hormones affect all insects with which they come in contact, including beneficial and innocuous species. Consequently, in Williams' words, "any reckless use of the materials on a large scale could constitute an ecological disaster of the first rank."

The second "sideshow" reads even more like something out of the laboratories of a modern Paracelsus. It is best told as the story unfolded. In 1964, Professor Williams invited Karel Sláma, of the Czechoslovak Academy of Sciences in Prague, to his laboratories at Harvard. Sláma brought with him cultures of the insect on which he had been doing much of his own research: Pyrrhocoris apterus, a small bug with a long sucking beak that it uses to feed not on blood but on the sap of linden trees. In the laboratory, it lives well on linden seeds and water; paper toweling is also placed in the cages to give the bugs something to crawl about on and hide in. In Prague, Dr. Sláma had reared these bugs through many generations without trouble. But in Cambridge, Massachusetts, all but one of 1,215 individuals, when they underwent their fifth and final molt, produced not adults but sixth-stage larvae or "adultoids" (imperfect adults). Only one produced a normal, sexually mature adult. Some of his sixth-stage larvae molted again to form even larger seventh-stage larvae or adultoids, but no additional adults. Was there something stultifying about the Harvard mystique; or had Williams simply been experimenting with juvenile hormone so long that his laboratory was heavily contaminated with it? Williams and Sláma compiled a list of fifteen differences in conditions in Cambridge and Prague, and began to test these one by one to isolate the cause of the trouble. (They did not, of course, attempt to compare the mystique of the two cities. Pyrrhocoris apterus has reddish wing covers, but it is hard to believe that it is greatly concerned with ideologies.)

Over a period of time, all but one of these fifteen possibilities was eliminated, and it was concluded that the paper toweling in the cages was responsible! When they replaced it with laboratory filter paper, all individuals developed normally. At first they thought there must be some chemical in the Scott toweling they were using that simulated the juvenile hormone, but further study showed that several other brands produced the same effect, and it made no difference whether the tissue were prepared for use in the kitchen, boudoir, or bathroom. As a matter of fact *The New York Times, Wall Street Journal, Boston Globe, Science,* and *Scientific American* were even more potent in inhibiting sexual maturity of the bugs. This was noted in an article in *The Times* on September 11, 1966, where it was pointed out that whatever the "paper factor" was, it might represent a powerful new means of insect control.

In the meantime Williams and Sláma found that the *London Times* and *Nature* (published in London) were inactive, as were a number of other paper materials from Europe. However, Japanese newspapers were as active as American ones (they later found out that Japanese newsprint is imported from Canada). They also found, curiously, that this "paper factor" had no effect on metamorphosis in the cecropia moth or any of the other silk moths in Williams' laboratories. Even more oddly, it had no effect on Rhodnius prolixus, a true bug related to Pyrrhocoris apterus. This led them to try the effect of cecropia hormone on Pyrrhocoris, but even though they injected large amounts into fifth instar larvae, all molted to produce normal adults—this in spite of the fact that cecropia hormone is fully effective on Rhodnius and many other insects. We think of hormones as being broadly effective on all animals of a given kind, but it appeared that Pyrrhocoris, though a fairly unremarkable bug in most respects, had its own juvenile hormone, which was mimicked by something in American (but not European) papers.

That there might be hormones specific to certain species was an important discovery. The danger of using hormones in control is that they affect all animals of that kind; use of juvenile hormone, for example, would likely eliminate all butterflies, honeybees, and so forth. But the use of "paper factor" might eliminate only the linden-feeding Pyrrhocoris apterus. Unfortunately, this isn't enough of a pest to make it worth the effort. However, Kailash Saxena, of the University of Delhi, India, has just recently discovered that the red cotton bug, a serious pest in India and North Africa, is also inhibited from reaching adulthood by "paper factor." Experiments are now underway in Delhi to determine whether the material can be used to control this insect effectively.

We now know that "paper factor" is a substance occurring in the balsam fir, a major source of wood pulp in North America. Workers at the United States Agricultural Research Service in Beltsville, Maryland, have identified the substance, and found its molecular structure to be somewhat similar to that of true JH. By the way, I assume the paper in this book contains "paper factor." If you don't like the book you might send it to India to control cotton bugs.

How does it happen that the balsam fir has a chemical substance that inhibits metamorphosis in certain plant-feeding bugs? It may, of course, be a biological coincidence. We believe that catnip is a naturally occurring plant substance that just happens to mimic one of the pheromones of the cat, so the idea is not without precedent. It is also possible, as Williams and Sláma have suggested, that the fir tree evolved this particular substance as a protection against certain plant-feeding bugs. The balsam fir now has no such bug attacking it, but the "paper factor" may represent a "biochemical memento" of the time when these trees did have an important natural enemy, which has since been rendered extinct or has gone on to attack some other tree. The idea of a plant evolving a chemical that repels or destroys plant-feeding animals is also by no means without precedent. After all, insecticides such as pyrethrum are of plant origin. Thomas Eisner, of Cornell, by the way, regards catnip as a substance that has been evolved by certain plants as an "insect repellent"; he has found that a great many kinds of insects flee when exposed to the vapors. Certain insects, such as walking sticks, produce chemicals very similar to catnip, which they forcibly eject to drive away natural enemies.

Plants produce a great many chemical substances the function of which we do not know. Is it possible that some of these were evolved as a protection against certain insects? In addition to the "paper factor" of balsam fir that mimics the juvenile hormone, recent research has revealed several plant chemicals that simulate the molting hormone ecdysone. Such substances have been isolated from ferns, evergreen trees, and weeds, and when injected into the bodies of insects produce abnormal molting. Indeed, some of them appear to be "superhormones"; a substance isolated from the roots of a Chinese weed was twenty-five times as potent as MH itself when tested on silkworms. These substances also occur in unusual concentrations. A few ounces of dried yew leaves are said to yield as much MH-like material as Peter Karlson was able to obtain from the four tons of silkworms he used in his original isolation of the hormone. Obviously this will prove a great boon to further research on insect hormones. But why should plants have evolved MH-like substances that have to be injected into the blood of insects to be effective and that (unlike "paper factor") have no effect when merely touched or fed upon?

As you can see, the study of the chemistry of living things is a veritable Pandora's box, opened just yesterday and filled with unimagined ideas we may someday put to use for our own good or ill. I have sketched the story of insect hormones in only its very broadest aspects; much could be added concerning the three hormones I have mentioned, and there are still others I haven't discussed. I have also said very little about the interaction of the various hormones with one another and with the nervous system and other body structures. To cover all these matters fully would require a couple of very fat volumes. But the volumes that may someday be written would fill several shelves.

Many persons have remarked on the resemblance of the endocrine glands and hormones of insects to those of vertebrate animals, including ourselves. We also have cells associated with our nervous system that release hormones (comparable to the brain hormone of insects). We also have a "master gland" just behind the brain, the pituitary, that secretes a hormone that promotes growth and other hormones that control reproduction. One of these hormones activates the thyroid gland, in the neck region, the hormones of which have many effects on body functioning and in tadpoles produce transformation to a frog. There is no point-by-point agreement in the structure and function of the glands, and the hormones are all quite different—after all, we are not bugs—but the surprising thing is that such utterly different creatures have evolved such basically similar systems of "chemical messengers" produced by glands in comparable parts of the body and interacting with one another and with body tissues in basically similar ways.

There are also many similarities in the research being done in the two fields. Various human illnesses and abnormalities have been traced to hormonal imbalance and can be treated accordingly, and even certain types of cancer can be ameliorated with the use of hormones. In some cases the hormones can be obtained from other vertebrate animals, and in other cases synthetic substances simulating hormones can be used. The search for "the pill" was largely a search for substances mimicking the hormones controlling the female sexual cycle and having no other important effects on the human body. Similarly, success in synthesizing substances simulating insect hormones from common chemicals such as farnesol, as well as knowledge of "paper factor," which is readily available in nature and which mimics the hormone of one group of insects only, has opened up a whole new field of research. In this case we hope the by-product will not be the insects' well-being but their selective demise.

There is little prospect that insect hormones will benefit man directly, even though some cosmetics manufacturers have already tried using "royal jelly" (a pheromone of the honeybee) in their face creams. Doubtless they are eyeing with interest the current work on the juvenile hormone—the "Peter Pan hormone," as Carroll Williams has facetiously called it. "Can we look forward to a chemotherapy for aging and senescence?" Williams asks. "Perhaps the juvenile hormone of insects may encourage a fresh look at this possibility." In the meantime, here is another theme for the science fictionists: Professor X drinks a potion of synthetic JH and turns into a youth—a youthful Rhodnius prolixus, that is.

To end on a more serious theme—I know of no field of research that better demonstrates the truly international character of scientific effort than the work on insect hormones. In this very incomplete survey I have mentioned an Englishman, an American, a Dutchman, a German, a Pole, a Czech, an Indian, and a Japanese. One scarcely thinks of these men as "an Englishman," and so forth—they are scientists, engaged in the one human effort that transcends language barriers, chauvinism, and all ideologies except the beauty of truth. Would that some of the vast energy frittered away on divisive and uncreative causes could be diverted to answering some of the very many questions still unanswered—and the answers to which lead inevitably to more questions.

Year of the Locust

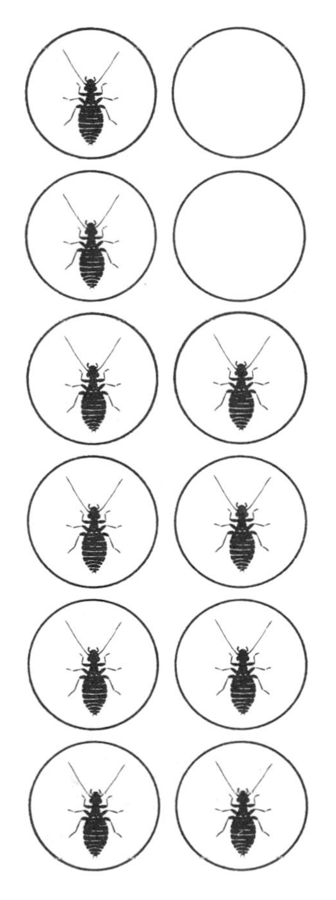

On a day in late July, 1874, the family of S. C. Bassett, of Buffalo County, Nebraska, gathered for supper. It had been a good day of hard work; it had been a good season, with just enough rain after several years of near-drought, and the crops were green and plentiful. Suddenly the sun was obscured by a dark cloud in the northwest: an odd time for a dust storm or a squall, but there it was. A few minutes later a member of the family went out to the well for some fresh water, but came back hurriedly with an empty pitcher and the startling report: "Grasshoppers!" As recounted by Everett Dick, in *The Sod-House Frontier:*

"The meal so happily begun was never finished. At first there was no thought of the destruction of the crops. All looked upon the insects with astonishment. They came like a driving snow in winter, filling the air, covering the earth, the buildings, the shocks of grain, and everything. . . . Their alighting on the roofs and sides of the houses sounded like a continuous hail storm. They alighted on trees in such great numbers that their weight broke off large limbs. . . . At times the insects were four to six inches deep on the ground and continued to alight for hours. Men were obliged to tie strings around their pants legs to keep the pests from crawling up their legs. . . . Men with clubs walked down the corn rows knocking the hoppers off, but on looking behind them they saw that the insects were as numerous as ever. . . ."

A few days later Freddy Funk, a lad of fourteen living on a farm in Marion County, Kansas, looked up from his chores to see a similar dark cloud approaching. As it grew suddenly darker the chickens fled to their roosts, and Freddy sensed an impending disaster. As he later recalled it for C. C. Howes, in *This Place Called Kansas:*

"Then with a whizzing, whirring sound the grasshoppers came from the northwest and in unbelievable numbers. They lit on everything. I was covered from head to foot. When they hit my face or hands the impact was like missiles and at once the insects began to eat. . . . We had about fifteen acres of corn which older settlers said would make fifty bushels an acre. The hoppers landed about four o'clock. By dark there wasn't a stalk of that field corn over a foot high left in the entire field. I slept in a straw stack. That night the hoppers ate my straw hat, or most of it, leaving me only a part of the brim and a part of the crown. They seemed to like sweaty things and ate around the sweatband of my hat. They gnawed the handles of pitchforks and other farm tools that had absorbed perspiration and they ate the harness on the horses or hanging in the barn."

On a day in mid-August, 1874, E. M. Correll, an attorney of Hebron, Nebraska, was returning home with his wife and baby in a surrey drawn by two ponies. A great cloud of grasshoppers approached from behind, and even though they urged on the ponies to the limit of their capacities, the insects soon engulfed them. They arrived home to find everything buried in grasshoppers. "Every vegetable, every weed and blade of grass bore its burden. On the clothes-line the hoppers were seated two and three deep; and upon the windlass rope which drew the bucket from the well they clung and entwined their bodies." Mrs. George Turner, of Fremont, Nebraska, came home in a swarm of grasshoppers "so thick that the horses could not find the barn." The Thomas Hewitt family, of Plum Creek, Nebraska, was deluged with the insects on the way to church. They listened to the sermon through a hail of hoppers on the roof, then drove home after the service to a farm they scarcely recognized:

"The cornfields looked as they would in December after the cattle had fed on them—not a green shred left. The asparagus stems, too, were equally bare. The onions were eaten down to the very roots. Of the whole garden there was, in fact, nothing left but a double petunia, which grandmother had put a tub over. . . ."

W. C. Thompson, editor of the *Larned* (Kansas) *Press,* went to his office on a Monday morning, leaving behind nine acres of corn and potatoes that "looked remarkably well and gave promise of a good yield. The grasshoppers came about 10 o'clock. At noon the besom of destruction was at work. At night the nine acres looked as though a seething fire had passed over it."

The noise of their coming was said to resemble "suppressed distant thunder" or "a train of cars in motion"; in a cornfield the noise of their chewing was said to sound like that of a great herd of cattle. The grasshoppers even ate the bark of trees, weathered wood from the sides of barns and houses, paper, freshly shorn wool, and dead animals—including dead grasshoppers. They entered houses and ate clothing and curtains. In the words of one contemporary:

"There is something weird and unearthly in their appearance, as in vast hosts they scale walls, housetops and fences, clambering over each other with a creaking, clashing noise. Sometimes they march in even regular lines, like hosts of pigmy cavalry, but generally rush over the ground in confused swarms. At times they rise high in the air, and circle round like gnats in the sunshine."

Many persons remarked on the fondness of the hoppers for tobacco and for onions. One farmer watched them devour his onions, then commented that as they flew past his house "their breath was rank with the odor of onions." This remark was in somewhat the same vein as that of another farmer who claimed that after stripping his tobacco to the stalks they gathered on his fence and "begged a chaw from everyone that passed." The editors of rural newspapers began to swap stories, perhaps in an effort to keep up the spirits of their readers. Thus, when an editor in Topeka, Kansas, reported that his picket fence had been partly eaten by grasshoppers, an Atchinson editor pointed out that his fence had been *entirely* consumed: and it was made of cast iron! Another reported that the hoppers got into his woodpile and cut it up into pieces "of convenient stove length."

A sense of humor was much needed. Most of the inhabitants of the affected areas were new immigrants, and life was hard enough even in good years—and the locusts had come on the heels of several years of drought. For they were locusts, akin to the Biblical locust and its relatives, that had so long plagued the Old World. Somehow the word "locust" had very early been misapplied in America; apparently some of the earliest settlers had seen swarms of the periodical cicada, and supposed they were locusts. The cicada is an insect that lives in the ground as a larva and that lives in trees and sucks sap as an adult: a very different insect from the true locust, which is simply a migratory grasshopper. But call them locusts, grasshoppers, or a scourge from heaven, their numbers were incalculable, their effects devastating beyond belief. Samuel Aughey, a teacher in Lincoln, Nebraska, reported that "a column of insects which from telegraph reports was at least 300 miles wide from east to west and averaging at least half a mile in depth, passed over Lincoln continuously from nine in the morning until three in the afternoon."

On November 16, 1874, a resident of Kearney, Nebraska, wrote as follows to the *Prairie Farmer* of Chicago, the leading agricultural newspaper of the day:

"The number of actual homestead settlers is . . . reduced fully one-half in my own neighborhood, and of that one-half, not one family in ten have provisions, fuel or clothing to last them through the winter. . . . There is no work, no money. There is no seed corn, and in very many instances, no seeds of any kind for another year's planting. On the 13th inst., I met two of my neighbors. One has a family of six to provide for, three of them young children. Says he: 'I have just flour enough to last until Saturday night.' The other has a family of ten, four of whom are sick. . . . This want extends over the whole area of country, west, north, and south, and the farther the settlement is from supplies, the greater the wants and privations of the settlers."

In Brown County, Minnesota, a widow wrote to her bishop: "All is lost. What to do I do not know. . . . I had a nice piece of barley ready to cut. There is nothing left but the straw, the heads lying thick on the ground. Dear Bishop, I am almost heart-broken, and nearly crazy, to think of the long, cold winter, and nothing to depend on. May God help us. . . ."

The sympathy of the nation was quickly aroused. Relief societies were organized in the East, and shipments of flour, corn meal, clothing, and dry goods were made to the Plains states. The legislatures of Kansas, Nebraska, and Dakota Territory voted to issue bonds to help the farmers, and the United States Congress voted $150,000 in goods to be distributed by the Army. Nevertheless, to individual families, some of them well isolated from roads and rail lines, help seemed slow in coming. Some of them lived for months on a few potatoes and other vegetables missed by the locusts—and on their chickens and turkeys, which had fattened on locusts to such an extent that they were said to taste like them. There were many scandals over the distribution of goods, and no small amount of graft and collusion. In Dakota public sentiment was against the relief bonds that had been voted, and nothing was done, while in Kansas the state supreme court declared the bonds unconstitutional. The year of the locust was indeed a dismal one for much of the Midwest, and there were those who wondered if the settlement of the plains had not been a mistake.

The *Junction City* (Kansas) *Union,* on August 22, 1874, reported six hundred wagons on the road, heading east, whence their occupants had come some years earlier in the search of a new life under broad and munificent western skies. But those skies had yielded far too little rain and far too many locusts. When a comet appeared in the skies that summer, it seemed to confirm the statements of those who claimed that this was a visitation from the Lord, an expression of wrath at the sin and corruption of the people. Their ministers read from Joel: "The day of the Lord cometh . . . The land is as the garden of Eden before them, and behind them a desolate wilderness: yea, and nothing shall escape them. . . ." And of course from Exodus:

"And the locusts went up over all the land of Egypt . . . they covered the face of the whole earth, so that the land was darkened; and they did eat every herb of the land, and all the fruit of the trees . . . and there remained not any green thing . . . in all the land of Egypt."

When, in the spring of 1875, it appeared that the progeny of the hordes of the previous year would shortly devour every green shoot in the fields, Governor Hardin of Missouri issued a proclamation, reading in part as follows:

"Wherefore, be it known that the 3rd day of June proximo is hereby appointed and set apart as a day of fasting and prayer, that the Almighty God may be invoked to remove from our midst those impending calamities, and to grant instead the blessings of abundance and plenty; and the people and all the officers of the State are hereby requested to desist, during that day, from their usual employments, and to assemble at their places of worship for humble and devout prayer, and to otherwise observe the day as one of fasting and prayer."

The people fasted and prayed as proclaimed by the governor, and lo, within a few days the locusts began to leave and to die; and by the fourth of July "the whole country presented a green and thrifty appearance again." The Lord had forgiven his contrite people. In retrospect, it appears that the governor had received, a few weeks before his proclamation, the annual report of the Missouri state entomologist, one C. V. Riley, in which the life history of the locust was discussed at some length and the prediction made that, since the locusts did not thrive permanently in the Mississippi Valley area, they would begin to leave Missouri in early June. In fact it was Riley himself, never an overly modest man, who was the first to point out the governor's fortunate timing. Two days after the governor's proclamation, Riley had written the *St. Louis Globe* that while he appreciated the sympathy of the governor for the suffering of the people:

"For my part, I would like to see the prayers of the people take on the substantial form of collections, made in churches throughout the State, for the benefit of the sufferers, and distributed by organized authority; or, what would be still better, the State authorities, if it is in their power, should offer a premium for every bushel of young locusts destroyed."

For this he was taken to task by the Reverend Dr. W. Pope Yeaman, of the Third Baptist Church of St. Louis, for taking "unnecessary pains to sneer at Providence." Riley in turn headed a section of his annual report "Not a Divine Visitation." The expression of the opinion that locusts were sent to punish people for their sins, said Riley, "is a downright insult to the hard-working, industrious and suffering farmers of the Western country, who certainly deserve no more to be thus visited by Divine wrath than the people of other parts of the State and country. Persons who promulgate such views are little removed in intelligence from the poor crack-brained negress whom I saw in the streets of Warrensburg shouting and imploring the people not to kill a locust, since God Almighty had sent them. . . ."

It will pay us to digress for a moment and look at this man Riley, for few more remarkable figures have ever crossed the stage of entomology. Indeed, there was something of the actor in Riley; he was strikingly handsome, with an elegant and well-groomed mustache, a swarthy complexion, and a full head of dark, curly hair. "There was always something unconventional and picturesque about his costume and appearance," wrote G. Brown Goode in a memorial appreciation in *Science*. "He looked like an artist or a musician, and indeed he possessed the artistic temperament in a high degree."

In 1874 Riley was only thirty-one, but he had already been state entomologist of Missouri for six years. In those times only a handful of states employed an entomologist, and it is to the credit of the state of Missouri that they had not only created such a position but had filled it with a young man of great energy who must have cut an unusual figure even in a burgeoning area where rugged individualists were by no means in short supply. Cut a figure in the annals of entomology Riley certainly did, for the nine annual reports he issued from his office in St. Louis excelled by far anything that had been done in this country up to that time, both in the quality of the research and in the very attractive manner in which it was presented. Riley's studies ranged far beyond the borders of Missouri, and the importance of his work was quickly recognized throughout the world. Even Charles Darwin was led to comment on Riley's powers of observation and to credit Riley with a great number of facts that he found valuable in his own work.

Charles Valentine Riley was born in Chelsea, London, and educated in France and Germany. At the age of seventeen he moved to the United States and went to work on a farm in Illinois operated by a man he had met in London. Riley had excelled in school, especially in art and in natural history, and he excelled in agriculture, but his health did not hold up under the rigors of the work of a stock farm. Urged by friends to find work less strenuous and more suited to his talents, he moved to Chicago, where, after a short stint of manual labor in a pork-packing plant, he managed to find a job as a reporter for the *Prairie Farmer.* Here his penchant for drawing and his enthusiasm for insects drew widespread attention. In particular he came to know Benjamin Dann Walsh, who had been a classmate of Darwin's at Cambridge University and who was then state entomologist of Illinois and one of the most eminent entomologists of the country. In 1868, Walsh and Riley founded a journal, *The American Entomologist,* and that same year Riley, at Walsh's recommendation, was appointed to the newly created position of state entomologist of Missouri.

A century has passed, and it is now hard to conceive how little was known of the insects of North America when Riley assumed his post. In 1841, Thaddeus W. Harris, the librarian of Harvard University, had published his classic *Report on the Insects of Massachusetts Injurious to Vegetation,* a well-illustrated compendium of knowledge about common insects and how to control them. (Picking them off by hand and crushing them was a common method; and at least we can say that in the ensuing century and a quarter no insect has developed resistance to this method!) B. D. Walsh and others had published important papers, but the amount of detailed information in Riley's Missouri reports far exceeded anything up to that time. That we no longer hear very much about some of the insects he was concerned with—the chinch bug and the grape phylloxera, for example—is in part attributable to Riley's reports and their effect on the subsequent blossoming of entomology in the United States. That the Rocky Mountain locust declined during his Missouri tenure, and was never again a major pest after 1878, was in fact a coincidence for which Riley could not and did not claim credit. Yet his name will always be thought of in connection with the decline of the locust because of the depth of his reporting and the case he made for government support of entomology.

In his report for the year 1874, Riley pointed out that this was by no means the first invasion by the Rocky Mountain locust. Study of old documents revealed that in the years 1818 through 1820, when few white men yet lived on the western plains, great hordes of locusts had been observed. Again in the years 1855 through 1857 they appeared, and again in 1864 through 1867. Riley found a common pattern in each invasion, and from this he was able to generalize about the origin and destination of the great swarms of 1874. On the basis of the reports of John C. Frémont and other early explorers, he concluded that the permanent home of the locust was in the high plains and plateaus of

Colorado, Wyoming, and Montana. Here the locust bred regularly, living upon buffalo grass and small herbs. During favorable years it built up very high populations, sometimes exhausting its food supply, particularly when the rains failed. In this case, "prompted by that most exigent law of hunger," the locust "rises in immense clouds in the air to seek for fresh pastures where it may stay its ravenous appetite. Borne along by the prevailing winds that sweep over these immense treeless plains

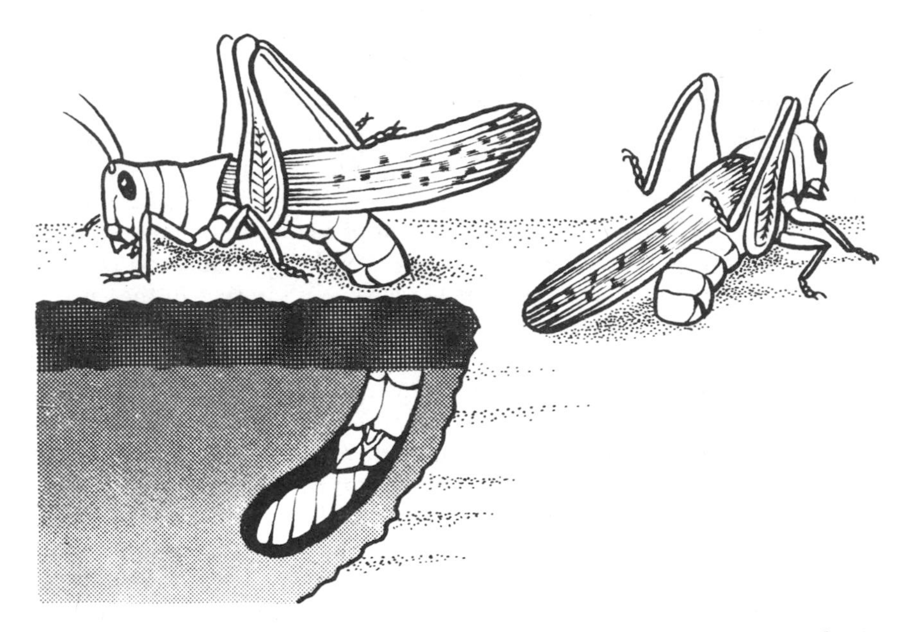

Rocky Mountain locusts laying their eggs. (From a drawing by C. V. Riley.)

from the northwest, often at a rate of 50 to 60 miles an hour, the darkening locust clouds are soon carried into the more moist and fertile country to the southeast, where, with sharpened appetites, they fall upon the crops like a plague and a blight."

Riley traced some of the swarms from Montana in mid-July to Texas in late September, a distance of some 1,500 miles in 75 days, averaging 20 miles per day. Although migrations from the "permanent area" were usually toward the southeast, the locusts emerging from eggs laid in the "temporary area," well east of the mountains, normally flew northward. The young hoppers, before acquiring wings, often "marched in armies" over the ground during the warm part of the day. Damage from grasshoppers bred in the eastern plains was usually temporary, as they often did not thrive here or soon migrated northward.

Knowledge that they will not last is, however, little consolation to a farmer who sees his crops devoured by locusts in a few hours. As Riley admitted, "man is powerless before the mighty host." Could nothing be done? Certain types of smudge fires could be used for protecting valuable fruit trees or other localized crops, but they were of little value unless the wind conditions were just right. Or one could plant crops that the locusts preferred not to eat: castor beans, for example. "Finally," suggested Riley, "in cases where . . . famine stares the people in the face, why should not these insects be made use of as food?" At the 1875 meetings of the American Association for the Advancement of Science, Riley presented a paper on this subject. He gave a number of recipes, several of which he had tried himself. He also told of preparing a locust supper at a hotel. The chefs could not be persuaded to handle them, so he was forced to do it himself, "with the aid of a brother naturalist and two intelligent ladies.

"It was most amusing to note how, as the rather savory and pleasant odor went up from the cooking dishes, the expression of horror and disgust gradually vanished from the faces of the curious lookers-on, and how, at last, the head cook—a stout and jolly negress—took part in the operations; how, when the different dishes were neatly served upon the table and were freely partaken of with evident relish and many expressions of surprise and satisfaction by the ladies and gentlemen interested, this same cook was actually induced to try them and soon grew eloquent in their favor; how, finally a prominent banker, as also one of the editors of the town joined in the meal. The soup soon vanished and banished silly prejudice; then cakes with batter enough to hold the locusts together disappeared and were pronounced good; then baked locusts with or without condiments; and when the meal was completed with dessert of baked locusts and honey *à la* John the Baptist, the opinion was unanimous that the distinguished prophet no longer deserved our sympathy, and that he had not fared badly on his diet in the wilderness."

Riley pointed out that many birds and mammals grow fat on locusts, as do certain Indians that use them regularly, so they are evidently nourishing. However, there is no evidence that he convinced many people. Rather, everyone with the slightest inventive mind set about devising some kind of contraption for catching or crushing them in large numbers. For example, a Mr. Michael Simpson (of Boston, no less) invented a wooden device covered with a hood, to be dragged by a horse for the purpose of scooping up and smashing locusts—the Simpson locust-crusher, it came to be called. The Adams locust-pan, on the other hand, had a reel in front somewhat like that of a harvesting machine, serving to scoop up the locusts and toss them into a long pan filled with oil. One of the most elaborate devices was invented by Mr. J. A. King, of Boulder, Colorado, and was known as the King suction-machine. This was a large tank mounted on a horse-drawn rig, and containing a fan that revolved with the wheels and created a suction that drew up the locusts through a vacuum-cleaner-like nozzle. The locusts collected by these and other types of machines were dumped into great heaps along the sides of the fields and on still nights were burned to make gigantic bonfires. All this did much to boost the morale of the beleaguered farmers, though in the final analysis it did little to diminish the hordes of locusts until after they had done their damage.

At Riley's urging, the states of Missouri and Minnesota enacted laws awarding bounties for the collection of locust eggs and hatchlings. In Missouri, a bounty of five dollars, half paid by the state and half by the county, was awarded for each bushel of eggs collected during the fall or winter. Hatchlings collected in March were valued at one dollar a bushel, the slightly larger hoppers in April at fifty cents, the still larger ones in May at only twenty-five cents a bushel. The locusts had to be presented to the county clerk, who then issued a certificate to be presented to the county treasurer. The office of county clerk was not an enviable one! It is said that on a few occasions grasshopper eggs were actually passed about as currency.

The Kansas legislature, in 1877, passed quite a different law, one that has become something of a classic in legal history. This is the so-called "grasshopper army" Act, which was apparently never enforced, although still on the books until 1923. This law required every able-bodied male between the ages of twelve and sixty-five to assemble for the purpose of fighting locusts, under the supervision of the road overseers, whenever so ordered by the town officials. Persons over eighteen could avoid duty by paying a dollar a day to the overseer, but anyone not paying or serving in the "grasshopper army" could be fined three dollars a day.

These various state laws were in large part the result of a conference of governors of the locust-ravaged states, held in Omaha in October, 1876. The governors were joined by several entomologists and agricultural experts, not the least of whom was C. V. Riley. The report of the conference appears to have been written in considerable part by Riley, for it urges state action of the type already outlined in Riley's reports. It also criticizes the United States Department of Agriculture for inaction, and calls for the appointment of a national commission to study the locust on a broad scale and to provide effective remedies. In his "Missouri report" of two years earlier, Riley had remarked, regarding the federal Department of Agriculture, that "the people have lost all hope of getting much good out of that institution." In the same report he pointed out that major insect pests, particularly of a migratory type, could never be dealt with effectively within the confines of one state and with the limited funds available. He therefore felt that the only solution was a national commission of entomologists.

Through the report of the Omaha conference, Riley's views came to the attention of Congress, and in the winter of 1876 two bills were introduced that would establish a three-man commission to study the Rocky Mountain locust and other pests of national importance. One bill would have made the commission responsible to the Department of Agriculture, the other to the Secretary of the Interior; but the bill that was finally passed, in March, 1877, made the commission responsible to the director of the United States Geological Survey! The creation of a new commission was not universally welcomed, as evidenced by an editorial in *The Nation* shortly after the bill had passed:

"In the future the Agricultural Commissioner will scatter the seed broadcast over the land, while the national entomologist will follow closely on his trail and exterminate the various bugs that may attack the ripening grain. We only want now another Commissioner to harvest the crops, and another to see that they get to deep water, and the husbandman will be entirely relieved from grinding toil."

Nevertheless the United States Entomological Commission became a reality, and went to work with a vengeance. The three entomologists were A. S. Packard, a student of Louis Agassiz who had authored the first American textbook of entomology; Cyrus Thomas, a Lutheran minister converted to entomology, and especially to grasshoppers; and of course C. V. Riley as chairman. The three divided the midwestern states between them, and each made a detailed inspection of his area and gave talks to various audiences, in which, in L. O. Howard's words, they "were able to bring relief to the burdened minds and to restore hope to the agricultural population which had virtually lost hope."

The first report of the commission, issued in 1878, was 477 pages long, with 27 appendices totaling 279 additional pages. The second report, only 322 pages long and with a mere 8 appendices, doubtless reflected the decline of the Rocky Mountain locust as a major pest. In 1878 the locust was less abundant even in its native home in the Rocky Mountain foothills, and the only swarms were small and local; and in 1879 hardly any at all were reported. The second report of the commission concluded on an encouraging note:

"A large proportion of the money losses resulting from the locust invasions . . . was the result of a panic, of uncertainty as to the future; this resulted in disheartenment, in the abandonment of large tracts of the best farming lands to nature and the locusts. This will probably never happen again in the West. The knowledge already disseminated, the extent of the population now pouring into the Northwest, the rapid settlement of the Territory of Montana, and the completion of the Northern Pacific, Canadian Pacific, the Utah and Northern Railroads, and the consequent change in the surface of the country due to human agency will so essentially modify the locust situation that we believe the West will never again suffer as in the past."

They were perfectly right. A small swarm was observed in 1888 in Minnesota, another in Nebraska in 1892. A few specimens were collected in North Dakota in 1900, a few more in Manitoba in 1901 and 1902. So far as known, no one has seen a single living specimen of the Rocky Mountain locust since 1902; the species, once so fantastically abundant, is evidently extinct! Despite exhaustive searching, it cannot be found even in remote valleys of the Rockies, though it is still possible that there is a small population somewhere that is capable of great increase under suitable conditions and that the species may yet again become the scourge of the prairies. Some have felt that there must be a connection between the decline of the bison and that of the locust. Possibly the destruction of the bison herds left an unusual amount of grass available, so that the locusts could build up great populations that later crashed of their own imbalance. However, it appears that bison were still very plentiful in the western plains and foothills in 1874, and they were surely plentiful during earlier locust plagues. The commissioners may well have been right that the settling of Colorado, Wyoming, and Montana had an important effect on the locusts. However, the sudden extinction of the species remains a riddle and probably always will. Grasshoppers of many species are still exceedingly abundant in parts of the West. In 1936, for example, various species were reported to have destroyed crops valued at over $100,000,000. But the Rocky Mountain locust was not among them.

Shortly before the turn of the century, prospectors in the high country of Montana, not far from Yellowstone Park, discovered a glacier that was unusual in having horizontal layers of a dark substance. Close examination showed that these layers consisted mostly of frozen Rocky Mountain locusts. At the lowest end of the glacier, where the ice was melting, there were great piles of rotting locusts. "Grasshopper Glacier" has since received considerable publicity, although evidently it has deteriorated somewhat in recent years. It is at an altitude of about 11,000 feet, not far from Granite Peak, the highest mountain in Montana.

Each layer in the glacier contained uncountable millions of locusts, and each probably represented one or more great migratory swarms that was carried by wind currents through the high saddle in which the glacier is located. Possibly the cooling effect caused the locusts to drop in great numbers onto the ice, or they may have encountered snow squalls that buried them immediately. That this may be the right explanation is suggested by the fact that even today one sometimes finds on the surface of the glacier quantities of live locusts belonging to a migratory species occurring chiefly in Nevada, but apparently carried and deposited here by air currents. It would be most interesting to know when these layers of Rocky Mountain locusts were deposited. In 1953, specimens were sent to Dr. Willard F. Libby, at the Institute of Nuclear Studies, University of Chicago, for dating by the carbon-14 method. His report indicated that the deposits were no more than a few hundred years old. Some of them may even represent parts of swarms that were reported by early explorers and settlers. In any event, we can say that there is no other extinct animal that has been so readily available in quick-frozen condition by the million!

The United States Entomological Commission went on to study other pest insects until its appropriations ran out in 1880. By this time a strong case had been made for federal support of entomology, and in 1878 Riley himself was asked to take over the position of chief entomologist of the United States Department of Agriculture. For nine months he held both this position and that of chairman of the commission, but it appears that he pushed too hard for greater appropriations for Agriculture, bypassing his superiors and going directly to Congress. This caused such a disturbance that he resigned his position in Agriculture, only to be reappointed two years later. Like many brilliant persons with high ideals and strong motivation, Riley tended at times to be unmindful of protocol or of the feelings of his subordinates. Although he was a skilled artist, he often let assistants make the greater part of a drawing; he would then add a few refinements and his initials. In his publications, he often failed to give full credit to his subordinates for their contributions to the finished product—a trait not at all rare in the annals of science.

As chief federal entomologist, Riley surrounded himself with a corps of capable and energetic workers, and within a short time was making many innovations in the work of the department. He began a series of bulletins of interest to farmers and agricultural scientists, and was influential in the passage of the Hatch Act of 1888, establishing state experiment stations. He continued the excellent studies of insect life histories he had begun in Missouri. For example, in 1892 he published an extended account of the remarkable association of the yucca moth with the yucca plant—research on an insect of no great importance to man, carried out entirely during his off-duty hours. Riley had been working on the yucca moth off and on for twenty years, and so sound was his research that his treatise remained the standard reference for seventy-five years!

Riley was also interested in the possibility of controlling pest insects by importing and releasing their own natural enemies, a procedure untried at this time. When the fluted scale insect threatened to wipe out the citrus industry in California, he requested funds to send workers to Australia, the native home of the insect, to look for natural enemies. When funds were refused, he persuaded the Secretary of State to send two men of his choice to Australia on an entirely different mission, secretly commissioning them to collect parasites of scale insects. The two men sent back a species of ladybird beetle that within a few years reduced the fluted scale to a pest of minor proportions, thus scoring a major triumph for Riley and the Department of Agriculture. As L. O. Howard, Riley's successor as chief entomologist put it, "obstacles as a rule only inspired Riley to further efforts."

Riley scored another great success with respect to the grape phylloxera, a small sucking insect that had been introduced into Europe from America and at one time had destroyed 2,500,000 acres of French vineyards. Since the insect was not overly serious in the eastern United States, Riley suggested that French grapes be grafted on to resistant American rootstocks. This suggestion was credited with saving the French wine industry, and for it Riley was awarded the cross of the Legion of Honor.

In 1894, Riley, although only fifty-one years of age, resigned his position because of near exhaustion from his many duties. He had earlier given his insect collection, numbering well over 100,000 specimens, to the United States National Museum, where it formed the nucleus of what is now one of the largest collections in the world. Upon resigning his position, he was given a room in the museum, where he hoped to devote his still considerable energies to some of the many problems he had never had time to work out properly. Unfortunately, on his way to work on the morning of September 14, 1895, his bicycle went out of control on a hill, struck a piece of curbstone at the bottom of the hill, and threw him head-first on the pavement. Within a few hours he was dead, and the meteoric career of a man who had come to the United States a penniless orphan thirty-five years earlier had come to an abrupt end.

With the death of C. V. Riley and the extinction of the Rocky Mountain locust, surely our story is ended. But it is not.

During 1895 the papers that reported Riley's death also reported widespread famine in the Congo and Cameroons—famine as a result of a great plague of locusts of a species much larger than the Rocky Mountain locust, the so-called "migratory locust" (a redundancy, since a locust is by definition a migratory grasshopper). In 1898 the locust plague extended into the Gold Coast, and in 1900 the Lake Chad area of central Africa was devasted by great swarms.

In 1898 a large part of Argentina was overrun by the South American locust, an even larger species. So serious was the situation that a law was passed similar to the "grasshopper army" laws of the American Midwest. All persons between the ages of fifteen and fifty were required to fight locusts in the area; they were paid a small salary, and were allowed to hire others to do their work or to pay a tax instead. The operation was under the direction of a governmental board called "Defensa Agricola," or "crop defense," which also supplied information and supplies to the farmers.

In 1906 a tremendous swarm of brown locusts emerged from the Kalahari Desert and descended upon the farms of South Africa. One swarm was said to be fifteen to twenty miles wide and sixty to seventy miles long. Behind them the ground was left covered with their droppings, and the veldt was stripped bare of anything green. The small hoppers that emerged from eggs laid by the migrants were termed by the Boers *Voetgangers*, or infantry, because they marched like a great army across the fields, leaving nothing edible behind.

In 1912 swarms of the migratory locust radiated out from the Danube delta into the Ukraine and into Bulgaria. A few years later some of them appeared as far west as France and England in considerable numbers.

In June, 1917, great swarms of locusts descended upon British Guiana. Every piece of cultivated land, even the plots of aborigines deep in the jungle, received its "detachment of winged locusts." Virtually all the corn and cassava in the country was stripped of its leaves, and it is said that even the coconut palms were sometimes defoliated.

In 1926 an outbreak of the desert locust began in the Sudan and Arabia, and in the following years swarms flew south into Kenya and other parts of East Africa and east into India and Turkestan. Damage in British territories during those years was said to exceed $3,000,000 annually. C. B. Williams, of the Rothamsted Experiment Station in England, was residing in Tanganyika at the time. He describes some of his experiences vividly in his book *Insect Migration:*

"[On January 29, 1929] we received a telephone warning shortly after breakfast that an immense swarm of locusts was passing in our direction. . . . An hour or so later the first outfliers began to appear—gigantic grasshoppers about six inches across the wings, and of a deep purple-brown. Minute by minute the numbers increased, like a brown mist over the tops of the trees. When they settled they changed the color of the forest. . . . The swarm was over a mile wide, over a hundred feet deep, and passed for about nine hours at a speed of about six miles per hour: at two insects per cubic yard (an under-estimate) it must have contained about ten thousand million locusts.

"Along the railway which runs from Moshi in Tanganyika to Voi in Kenya, through the arid bush country round the foot of Kilimanjaro, the locusts settled in thousands, and seemed to prefer the railway line to the apparently equally attractive land on either side. If a swarm settled on the line where there was a slight up-gradient, the driving wheels of the engine slipped on the fat which oozed from the crushed bodies, and the train came to a standstill. One of my jobs was to see how this could be prevented; and I shall ever be grateful to the locusts—if for nothing else—for providing me with an excuse to travel for miles lying full length on the cowcatcher of a railway engine. Who can say that the life of an entomologist is dull?"

During these same years the migratory locust was building up huge populations in the basin of the river Niger, and by 1930 swarms had spread to the Sudan and thence southward into countries still reeling from the depredations of the desert locust. Meanwhile another species, the red locust, was expanding over much of Africa south of the equator, and in the mid-1930's the government of South Africa spent over $2,500,000 on defense of crops against this locust.

In the 1920's locusts were such a plague in Mexico that citizens were required to use a "locust stamp" on all letters, in addition to regular postage, in order to help finance the campaign against the pests. In 1934 Mexico and Guatemala signed a joint agreement to fight the locust, and later other Central American countries were invited to join. In the meantime, in 1933, damage by locusts in Argentina reached a new high, and a few years later the grasshopper army was replaced by "anti-locust preventive police," a corps of highly trained scientists and technicians using the most advanced reconnaissance and control techniques.

In 1939 an unusual infestation of grasshoppers appeared in the Big Smoky Valley of Nevada, and in the ensuing years migratory swarms spread into Oregon and northern California. These insects proved to belong to a previously unrecognized species, which was duly named by Ashley B. Gurney, grasshopper expert of the United States Department of Agriculture. In 1949 and again in 1957 a different species with migratory habits, called, appropriately, the "devastating grasshopper," did extensive damage to rangeland in California. In 1956, Frank T. Cowan, of the United States Department of Agriculture Station in Bozeman, Montana, reported to the Tenth International Congress of Entomology that some sixty species of grasshoppers, most of them solitary or only slightly migratory, cause damage to crops and rangeland west of the Mississippi estimated to average $75,000,000 per year. This estimate was in part based on the observation that grasshoppers consume annually about 6 to 12 per cent of the available forage—that is, between 5 and 10 million tons—enough to support livestock worth many millions of dollars. The Rocky Mountain locust must have snickered from his grave.

These accounts of depredations could be prolonged indefinitely. The Australian plague locust caused great damage from 1935 to 1939. In 1954–1955 a serious plague in Morocco is estimated to have cost that country over $10,000,000. In 1962 an invasion of Pakistan and India by the desert locust was so serious that those countries called on their armies and air forces to assist in combating them. The fact is that every year is the "year of the locust" somewhere in the world, and it is frightening to realize that in spite of man's remarkable technological advances these insects are still capable of building up fantastic populations in certain areas and then fanning out to destroy rich farm- and rangelands.

Man's unpleasant associations with locusts can, of course, be traced deep into antiquity. We not only have vivid accounts of locust plagues in the Bible, but ancient Assyrian and Egyptian records show that in much earlier times man was much concerned with locusts. C. B. Williams, in his book *Insect Migration*, reproduces figures from the VI Dynasty at Saqqarah, Egypt (about 2350 B.C.), showing locusts feeding on vegetation. The importance of locusts to the Aztecs in Mexico is well shown by the widespread use of locust figures in their art; and these insects have been immortalized in the name of Mexico City's great park, Chapultepec ("home of the grasshopper"). But in the twentieth century A.D., when man has carried his technology and his yearnings to the farthest corners of the earth—and indeed into space—the locust can no longer be tolerated as a companion. Why has it taken us so long to come to terms with the locust? For the very same reason we are frustrated by many of the earth's other "lesser creatures": we do not understand them.

But we are making progress, and as usual the locust turns out to be a pretty complicated animal, no matter how much he may look like some kind of a mechanical eating-machine mass-produced by demons. If all that has been published on locusts were stacked in one pile, it would be quite a pile—enough to feed a locust swarm for several hours, no doubt. Nevertheless we are just beginning to find out what makes them tick; and what we have found out is quite a story.

The first breakthrough was made by a taxonomist, one of those underrated denizens of museums and university backrooms who quite often come up with insights that prove valuable to those in more fashionable fields of biology. This particular taxonomist, B. P. Uvarov, did indeed receive recognition for his discoveries. Now Sir Boris Uvarov, he described his initial studies in his 1961 presidential address to the Royal Entomological Society of London in these words:

"It was 50 years ago that I was first faced with a problem concerning locusts which appeared to me to require only a short study by a taxonomist to be resolved. As I had by then two incomplete years of experience in the systematics of [grasshoppers], there was no doubt in my mind that I was fully competent to deal with this little problem. The problem was the existence of the two supposed species of the migratory locust . . . which differed strikingly in their [structural] and colour characters, as well as in their habits, the first occurring in swarms, the second singly; they were, however, vaguely connected by individuals and populations with intermediate characteristics. Of course, I thought, this was due to other entomologists not being critical enough to be able to find the real distinguishing characters between the two species, and I set out to do so. . . . However, further study . . . brought nothing but disappointment, since no single character, nor any combination of them, served to separate the two insects and my preconceived ideas were undermined. The final enlightenment was due to the locusts themselves, when from the eggs laid by swarms of one supposed species in 1912 there arose next spring a progeny which included a mixture of the swarming type individuals with many intermediate ones, as well as some typical of a non-swarming population."

The idea gradually came to Uvarov that this species existed in two forms capable of transforming one into the other, the intermediates representing transitional individuals. The two extreme forms, or "phases," as Uvarov termed them, differed in several ways: members of the gregarious phase were notably darker than those of the solitary phase, and had shorter hind legs, longer wings, and a lower and narrower thoracic shield just behind the head; they were also much more restless, with a tendency to "march" and to migrate. Uvarov postulated that the solitary, sedentary phase occupied a permanent breeding ground and from time to time produced large numbers of the gregarious phase, which then swarmed into new regions, where solitaries were once again produced. Thus, in his words, "the periodicity of locust invasions is caused entirely by the wonderful phenomenon of the transformation of a swarming locust into a solitary, harmless grasshopper."

Uvarov's original work was done in his native country, Russia, but at the time of the Bolshevik Revolution he was forced to leave, and he migrated to England, where his phase theory was finally published in 1921. In the meantime he had corresponded with Jacobus Faure of the

University of Pretoria, South Africa, who had discovered that the brown locust of that country also had a solitary and a migratory phase, the two differing in much the same way as the phases of the migratory locust studied by Uvarov. A few years later, Professor Faure showed that by bringing in young solitary-phase hoppers from the field and rearing them under crowded conditions in the laboratory he was able to produce variation in the direction of the gregarious phase; moreover, the progeny of locusts reared under crowded conditions, if also reared crowded, were wholly converted to the coloration of the gregarious phase in two generations. Within a few years it was discovered that several other species of locusts also exhibited solitary and gregarious phases. In 1928 Uvarov produced a large and attractive compendium: *Locusts and Grasshoppers: A Handbook for Their Study and Control.* This book was published by the Imperial Bureau of Entomology of London, where Uvarov held the position of senior assistant, and proved a most valuable summary and basis for further work. In this book, Uvarov was able to say that "numerous new facts permit it to be stated definitely that the phases of locusts are *a fact* of great biological significance that supplies a key to the problem of periodicity."

In his book, incidentally, Uvarov pointed out that C. V. Riley had suggested in 1877 that the Rocky Mountain locust might, after swarming, "degenerate" in the direction of a common Midwestern grasshopper very similar to it, the so-called "lesser migratory grasshopper." After the appearance of Uvarov's phase theory, several American entomologists proposed that the Rocky Mountain locust might merely represent the gregarious phase of the lesser migratory grasshopper and that conditions had simply not been suitable in recent years for the production of that phase. In 1932 Professor Faure was invited to come to the United States and conduct breeding experiments with the lesser migratory grasshopper similar to those he had conducted earlier on the brown locust. His work, conducted at the University of Minnesota, was unfortunately limited as to time, and he had difficulty obtaining large numbers of grasshoppers of the correct species. He did find that immature lesser migratory grasshoppers, when reared under crowded conditions, assumed darker colors, approximating those of the Rocky Mountain locust, and he concluded from this that the two were phases of one species. But we now know that quite a number of insects, when reared under crowded conditions, become darker, apparently as a result of complex factors affecting the amount of dark pigment deposited in the body wall. Faure himself was the first to point out that "army worms," the larvae of certain moths, become distinctly darker when kept crowded. We also know that the penis of the male Rocky Mountain locust had a peculiar twisted hook not present in any other member of its group, and it is too much to suppose that this was merely a phase difference. Quite likely the Rocky Mountain locust had a solitary phase, for its swarming behavior was very much like that of the migratory and brown locusts. Evidently the solitary phase is also extinct.

But to return to Uvarov. His book, as I mentioned, appeared in 1928. By coincidence, that was the year when the desert locust outbreak reached serious proportions in countries of East Africa then under British administration, to be followed shortly by new plagues of the migratory and red locusts. In 1929 a "locust committee" was established in Britain to see what could be done to find the causes of periodic swarming. The committee procured an appropriation of the grand sum of $1,400 to enable the Imperial Bureau of Entomology to start an information center relating to locusts. Uvarov was, of course, the logical man to head this center, and he did so. By 1931 the international aspects of the locust problem had become so serious—locust swarms being unconcerned as they are with political boundaries—that an International Locust Conference was assembled in Rome. At this conference it was decided to accept the London center as an international organ for the collection of information of all kinds on locusts. In 1945 the center became officially known as the Anti-Locust Research Centre. (Why not simply Locust Research Centre? Presumably so no one in Parliament would get the idea they were biased in favor of the locust.) In 1964 the Centre moved to new and expanded quarters containing a great variety of equipment of value in studying the behavior and control of locusts, as well as space to house its extensive records and library. The Centre now has a staff of more than sixty scientists, as well as facilities for visiting workers and students in training. Sir Boris Uvarov retired as director in 1959, but as of this writing he is still active, and in 1966 he published the first of two volumes of a new compendium of knowledge on the subject, titled (what else?): *Grasshoppers and Locusts.*

The Anti-Locust Research Centre, though paradoxically located in a country that has no locusts, has become a major research and information center for the world. Within recent years locust research and control have taken on a more and more international character. There are many examples of cooperation between neighboring countries, and International Locust Conferences have continued to be held periodically. Recently the Food and Agriculture Organization of the United Nations (FAO) has entered the picture, trying to plan a "global strategy of attack" against the locust. This recent mustering of human forces against these insects will very likely pay off, as man is rarely stymied indefinitely by any animal (except, of course, himself). In any event, some of the recent research on locusts has led us down fascinating and unexpected channels.

Uvarov's phase theory is still fundamental to locust studies, though with a number of modifications and refinements. In a review of the subject in 1950, the Australian entomologist K. H. L. Key remarked that "the fundamental cause of outbreaks of locusts is multiplication, and the 'irregular periodicity' of outbreaks is primarily due to changes in the population in the outbreak areas. . . . Phase transformation must be regarded as a secondary concomitant in the development of locust outbreaks. . . ." What environmental factors favor the population explosions of locusts? Precisely how does crowding produce darker hoppers that grow to be adults with relatively shorter legs and longer wings? What is the relationship between these changes in color and structure and the restless, migratory disposition of the gregarious phase? Do migrants deliberately set off in a particular direction, and if so, how do they navigate over terrain they have never seen before? Since swarming populations generally decline markedly after a few months or at most a very few years, of what ultimate benefit is swarming to a locust species? These are a few of the thorny problems toward which recent research has been directed.

Jacobus Faure described incipient swarm formation as follows: "A certain number of solitary type hoppers occurs on a given area. If the population is dense enough, some of the hoppers come close enough together to stimulate one another into jumping activity, and at the same time their strong tendency to gregariousness begins to assert itself and tends to keep them together. . . . The little group . . . is bound to capture more recruits, and this continues until a loose swarm is formed." Already some individuals will exhibit the darker patterns of the gregarious phase, and if crowding continues the next generation may be expected to contain large numbers of the gregarious phase, finally resulting in a crop of long-winged adults that take off en masse to another region. In the brown locust studied by Faure, incidentally, males of the gregarious phase are more than 1.5 times the size of the male solitaries, although the female size difference is not nearly this great. The immature solitary hoppers, in this species as in several others, have colors that are rather variable but that tend to blend with the background; but when, in Faure's words, "the progeny of these scattered locusts [appear] in swarms six months later there [is] no trace of such a protective resemblance; they all [wear] the King's regulation swarm uniform": in this case black with orange markings.

Faure postulated that as a result of their high rate of activity, gregarious hoppers produce "a surfeit of excretory products," some of which are deposited as dark pigments in the body wall. He called these excretory substances "locustine." He considered it probable that since the blood of adults was "more or less saturated with locustine," the yolk of the egg also contained locustine, and the young hatchlings were therefore darker and more active, their activity tending to build up more locustinc, and so on in a vicious cyclc. This explanation made it unnecessary to believe in the inheritance of features acquired during an individual's lifetime, a theory popular in recent years only among the Russians.

Many of these ideas are amenable to experimentation beyond that attempted by Professor Faure, and several workers at the Anti-Locust Research Centre have addressed themselves to these problems. Peggy Ellis, in particular, has refined Faure's techniques in various ways so as to isolate the factors involved in producing increased gregariousness. She has worked principally with the migratory locust, and primarily with the hoppers (that is, the immature, wingless stages). Early in her work, she noted a tendency for the hoppers to gather in the corners of their cages, so she devised various kinds of circular, cornerless cages which eliminated that cause of clumping. One such cage, lighted from above, had a floor divided by lines into thirty sections. Fifteen hoppers were introduced at a time and their positions recorded after they had settled down: rather like a roulette wheel, except that the wheel was motionless and fifteen six-legged objects took their own places within it. By the laws of chance, some 35 per cent of the hoppers, on the average, would end up two or more to a section, and this is essentially what she obtained when she tested two species of nonswarming grasshoppers in the apparatus. Hoppers of the migratory locust, when previously reared in isolation, clustered hardly any more than this, but hoppers of this species that had been "reared crowded" showed a strong tendency to cluster—50 to 70 per cent of them ended up two or more to a section. Evidently crowded hoppers get used to the presence of close neighbors—that is, they learn to aggregate. Dr. Ellis' isolated hoppers, when reared crowded for only four days, showed a strong tendency to aggregate; but her crowded hoppers, when placed in "solitary confinement" for four days, showed only a slightly reduced tendency to "socialize"—suggesting that this learning is largely irreversible.

In a second series of experiments, Dr. Ellis introduced hoppers into a circular cage in which she had placed three kinds of decoys: tethered live hoppers, pieces of white fiberboard, and pieces of black fiberboard. Hoppers reared crowded generally aggregated with the tethered live hoppers. Those reared in isolation tended not to aggregate with any of the decoys, but when they did so they preferred the black pieces of fiberboard. In a situation such as this, one is always tempted to draw analogies with humans: the child reared with few social contacts who grows up to be an individualist versus the child taught the "social graces" from an early age who grows up to be a Rotarian. The saying "It takes all kinds to make a world" may apply equally well to the world of humans and to that of locusts.

Is the sight of other locusts important in producing more "sociable" locusts? When Dr. Ellis placed individual hoppers in small celluloid cages inside a large cage full of locusts, the effect was to increase the subsequent tendency of the individual hoppers to aggregate only slightly. On the other hand, single hoppers placed directly in a crowd of nongregarious grasshoppers of another species *did* learn to aggregate, even though the grasshoppers did not. They even learned to aggregate

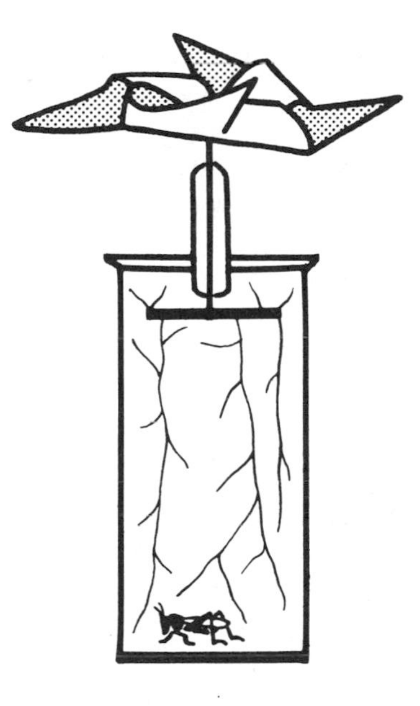

Peggy Ellis' "locust tickling machine." When the pinwheel is placed in the draft from a fan, the disc rotates and the locust is repeatedly touched by the ends of the fine wires. Such locusts, when removed from the apparatus, show a strong tendency to aggregate with other locusts.

when placed in a crowd of sowbugs, which are terrestrial crustaceans, and not even insects! Evidently touch is all-important, but it does not have to be the touch of another animal of the same kind. Will an inanimate touch stimulus suffice? To answer this, Dr. Ellis invented a "locust tickling machine" consisting of a jar containing many fine wires suspended from a circular disk that was rotated by a wind vane. She found that locusts placed individually in this apparatus for seven hours, and thus touched continuously in the absence of other hoppers, learned to aggregate as well as hoppers kept crowded for an equal period. However, hoppers merely forced to exercise for seven hours, in isolation, showed no increased tendency to aggregate. Dr. Ellis then went on to ask further questions:

"If learning to group in locusts involves an habituation to being touched, then we must answer the question: what holds them together, under natural conditions, long enough for them to become conditioned to one another? The habitat of locusts is always a patchy one and other

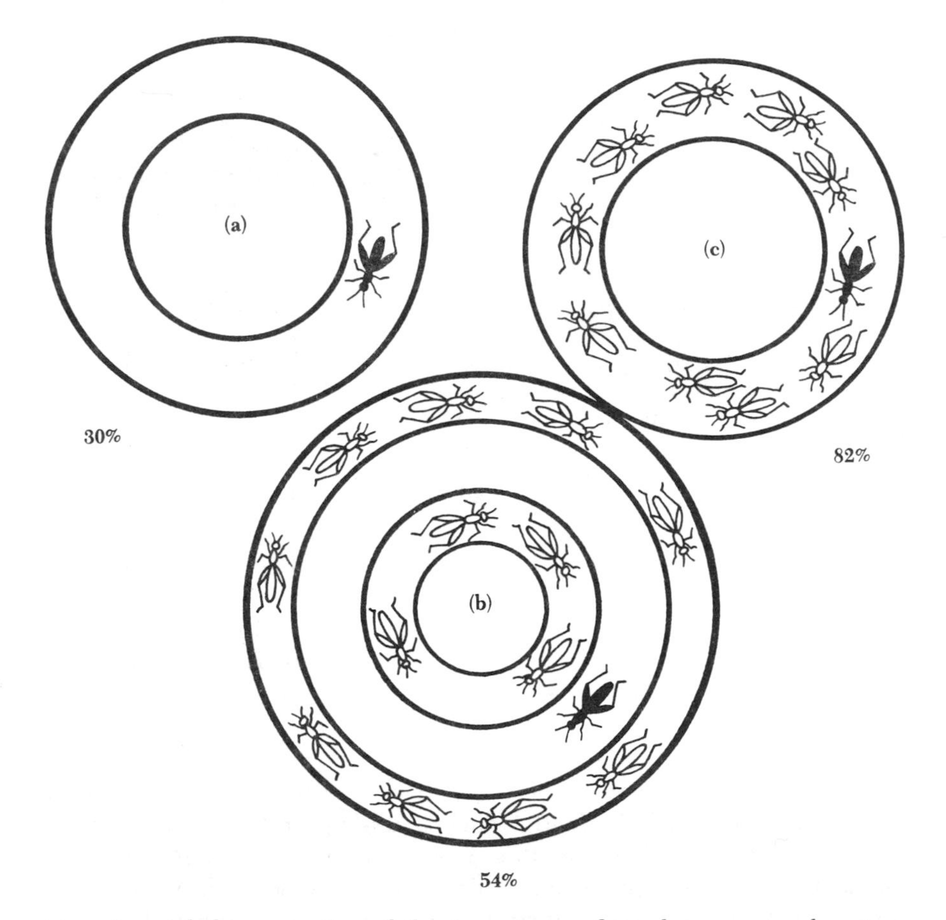

Experimental demonstration of the importance of touch in causing locusts to march. A test locust (*shown in black*), which has been reared under crowded conditions, marches only 30 per cent of the time when alone (*a*). In a cage in which he is able to see other locusts marching (*b*), he marches 54 per cent of the time. But when he is able both to see and to touch other marching locusts (*c*), he marches 82 per cent of the time. (After Peggy Ellis.)

workers have already suggested that the individuals meet when many come together in the restricted areas most suitable for basking, that is, resting in the sun, or when they come together to feed from scattered plant tufts."

To test the hypothesis that hoppers may learn to aggregate when brought together in basking groups, she returned to her earlier experiments using live, tethered decoys in a circular, evenly heated cage. Hoppers reared in isolation, you will recall, spent little or no time near the decoys, while hoppers reared crowded readily aggregated with them. When she placed small heating elements under each decoy, however, the hoppers reared crowded showed only a slightly increased tendency to aggregate, while the hoppers that had been reared in isolation showed a much greater tendency to clump in the heated areas with the decoys. When these were then transferred, after several hours, to evenly heated cages, they clumped with the decoys nearly as well as hoppers reared under crowded conditions from the beginning.

Evidently the well-known tendency of hoppers to bask in sunny spots may, when populations are sufficiently high, be sufficient to bring about the repeated touch stimuli necessary to "teach them to aggregate." Once hoppers have become conditioned to living in bands, still another type of behavior begins: the bands begin to "march," that is, to move over the ground in a concerted, continually moving horde. During marching, touch stimuli continue to keep the hoppers together. But at this time visual stimuli are evidently somewhat more important. Dr. Ellis used ring-shaped cages to demonstrate that a hopper reared crowded marches, on the average, only 30 per cent of the time when alone; when in visual but not touch contact with marching hoppers, such an individual marches about 54 per cent of its time; but when in both visual and touch contact with other hoppers, it spends most of its time marching. Exercising isolated hoppers in a treadmill improves marching performance slightly, but not a great deal. One other interesting item: other things being equal, hoppers that are the offspring of parents reared under crowded conditions are better marchers than those of parents reared in isolation. However, parental rearing conditions seem to have little to do with initial tendencies to aggregate. In spite of this, aggregating is a prerequisite to marching; a solitary-reared hopper, when dropped into a marching band, will simply not get into the swing of things—a draft-dodger, if you like.

If you are slightly confused by all this, it only proves how far from being simple automata locusts are. Their behavior is influenced by many different factors, and in subtle and complex ways. We should remember, too, that hoppers kept together long enough begin to show color changes, from the green and gray protective colors of solitaries to the black and orange of the gregarious phase. Their hind legs show a decrease in length, relative to other body parts, at successive molts, and when they reach the adult stage they have proportionally longer wings. How can the fact that they have "learned" to aggregate and to march possibly influence their body form and color? We now know that the word "learning" is rather misplaced, at least with respect to marching. David Carlisle and Peggy Ellis of the Anti-Locust Research Centre found that in the hoppers reared alone, the thoracic glands were consistently larger than they were in hoppers reared under crowded conditions. When parts of the glands were excised from solitary hoppers, they proceeded to acquire darker colors at the next molt. When gregarious marching hoppers were injected with extracts of thoracic glands, thus increasing the level of the hormone secreted by these glands, they showed much less tendency to march for several hours after the operation. Operations of this type are, of course, always accompanied by "controls," that is, individuals with "dummy operations"; otherwise we could not rule out the possibility that the operation alone caused the observed effects. Apparently the various stimuli provided by other locusts in close proximity bring about a recession in the activity of the thoracic glands—not exactly learning in the usual sense. Whether these glands produce more than one hormone or whether the "molting hormone" has more than one effect we do not know.

At the University of Strasbourg, in France, Professor and Mme. Pierre Joly have meanwhile been tampering with the other major endocrine gland of locusts, the corpus allatum. They implanted corpus allatum substance into gregarious hoppers approaching maturity, and obtained adults which varied in the direction of the solitary phase; they were also able to produce green, "solitary-colored" hoppers by increasing the amount of corpus allatum hormone in the blood of young, dark-colored, gregarious hoppers. This seems to contradict the results of Carlisle and Ellis, who found that it was an increase in the secretions of the thoracic glands that made hoppers "more solitary." Obviously there is much we still do not understand; the effects of the two hormones are doubtless not antagonistic, but occur at slightly different times and act upon different elements in the nervous system and growth centers. Once again the answer to one question has opened up a host of other problems.

Another facet of the problem still incompletely understood is the manner in which a hopper inherits a greater tendency to march from crowded parents than from isolated parents. The matter is a critical one, since we know that it takes two or three generations to produce adult locusts that are "fully gregarious" and ready to take off in a mass flight; yet at the same time all other evidence teaches us to reject the inheritance of acquired characters. In the words of J. S. Kennedy, of the British Agricultural Research Council: "It is widely held that no conceivable mechanism exists by which acquired habits could be transmitted through the egg, although locusts apparently have one."

Current feeling seems to be that this is a form of "cultural inheritance." Crowded female parents, presumably as a result of hormonal factors, lay fewer and larger eggs. These eggs produce offspring that are larger, darker, and more readily conditioned to aggregate; furthermore they have more fat, vitamin C, and water in their bodies, and are able to live for some time if food is not readily available. Offspring of crowded parents develop more rapidly than "solitaries," and require one less molt to reach maturity. Is all this simply the result of starting life under crowded conditions and from a large egg with more yolk than usual, or is some specific chemical, hormonal or otherwise, transmitted through the egg? The search for a solution to this problem is likely to entertain locust experts for some years to come.

This brings us to still another remarkable feature of locust development. Groups of desert locusts tend to be synchronized so that all are ready to mate and lay eggs at one time—a requisite feature of a good migrant, since the whole of a huge population has to be ready to swarm at about the same time. The male desert locust, when sexually mature, develops a bright yellow coloration, and it has been observed that the presence of one or more of these yellow males in a group of locusts of varying age causes the younger ones to speed up their development. Werner Loher, of the University of Tübingen, Germany, discovered that these yellow males produce a pheromone over their body surface and that this pheromone is picked up by other locusts either by rubbing against the mature male or by smelling the substance with their antennae. An elaborate series of experiments showed without a doubt that prolonged exposure to this pheromone hastened maturation and tended to synchronize the swarm. Smears of the substance on pieces of cotton served as well as the locust itself, and blinded locusts reacted as well as unblinded ones, indicating that the sight of the mature yellow male has nothing to do with it. Dr. Loher found that males that had had their corpus allatum removed failed to develop the yellow coloration or the pheromone, but when these glands were implanted into such individuals they proceeded to turn yellow and produce the pheromone. It would be hard to find a better example of the intimate coordination of hormones, pheromones, behavior, development, and structure.

There is still more to this wonderful story. It is well known that some plants produce gaseous substances that serve to synchronize the growth of all plants in that particular group. Ripening apples, for example, give off ethylene in small amounts, and this hastens ripening, so that apples in storage all tend to ripen at the same time. In the laboratory, the same effect can be produced by painting a similar substance, chloroethanol, onto the apples. Drs. Carlisle and Ellis have recently found that this substance has the same effect on the desert locust. They painted sixteen locusts with chloroethanol dissolved in alcohol two days after they reached adulthood, and a second group of sixteen with plain alcohol. The two lots were equally crowded, and at the end of nineteen days the sixteen treated with chloroethanol were appreciably closer to the coloration associated with sexual maturity; many of them had mated and some were nearly ready to lay eggs. On the other hand, those treated only with alcohol had barely begun to mate, and the females contained much smaller, undeveloped eggs. It is entirely possible that in nature similar plant substances are important in bringing about the synchronization of locusts within a given area, and hence "preparing them" for swarming. Indeed, Carlisle and Ellis found that when jars of myrrh (an aromatic herb of Biblical fame) were placed in cages of locusts, sexual maturation was hastened.

We now believe that the population build-ups necessary to start the locusts on the road to phase transformation and ultimately to migration are the result of combinations of climatic factors suitable for the survival and reproduction of both the locusts and the vegetation on which they feed. Sooner or later, often as a result of drought, the vegetation gradually deteriorates in parts of the habitat and finally over much of the habitat. This tends to crowd the great masses of hoppers into more limited places, and finally, synchronized as they are, winged adults arise in a great cloud and disappear from the scene. Swarming is, then, a safety valve, a means of avoiding a great population crash. As J. S. Kennedy has pointed out, the swarming of locusts should not be thought of as a "negative, suicidal act." The migrant swarms do eventually decline and disappear, but not without inoculating new areas with locusts. Locusts left behind in the original area or at various way stations revert to the solitary phase, and in some cases even the solitaries fail to survive in the new situations. But when and where the situation is right for another population explosion, the locusts are there to take advantage of it. One thinks of the suggestion made in Chapter 1, that man may eventually use space travel as a safety valve for overexploitation of the earth, sending out periodic "swarms" into space in the hope that some will find new places suitable for human life.

Kennedy conceives of locusts as existing in two forms differing not only in color, structure, and behavior but also in ultimate survival value for the species (not unlike the two or more forms of mimetic butterflies, but without a hereditary basis). He speaks of the "division of labor between the phases": the solitary phase is "vegetative"; that is, it exploits readily available food resources and converts these efficiently into more locusts; on the other hand, the gregarious phase specializes in dispersal and the finding of new food resources. It is true that adult "solitaries" do sometimes leave exhausted habitats, but generally in diffuse swarms in the evening, rather than by day; and they travel only until suitable food is located. True migrating swarms of the gregarious phase, however, may eventually travel hundreds of miles, and may descend upon sites where food is relatively sparse and readily consumed within a short time. Swarms of desert locusts have been observed in the Atlantic more than 1,200 miles from land.

In general, locust swarms do often seem to emanate from drought-stricken areas, and they often seem to end up in areas where rain is falling. Desert locusts, for example, can survive for long periods in areas of extreme aridity, but they require succulent vegetation for maturation. Swarms have a way of leaving localities in the early part of the dry season and showing up in another area near the start of the local rainy season. Generally, the desert locust spends the winter in a belt subject to winter rains, running through North Africa, the Middle East, and Pakistan; but as summer approaches swarms move hundreds of miles into areas subject to summer monsoon rains, chiefly south of the Sahara and Arabian deserts and in the Indian peninsula. Since these represent successive generations, how do the locusts "know" which way to migrate, and how do they guide themselves in the proper direction?

A great deal has been written on this subject, some from comfortable armchairs and some from notes jotted down in an airplane in actual pursuit of a swarm. It appears that mass flights usually begin in the morning. Basking groups first collect in sunny places, and mutual stimulation soon results in much milling about and some jumping into the air, followed by a few disorganized flight movements. Eventually a whole basking group takes wing, joins with other such groups, and finally the churning, chaotic swarm begins to move off in a common direction, usually into the prevailing wind. It has been shown that locusts can maintain an airspeed of about ten miles an hour for more than twenty hours. However, as the swarm rises from the ground it soon encounters air strata where the wind velocity is greater than this. Individual locusts may be oriented crosswind, or their direction may vary from time to time, those on the periphery of the swarm having a tendency to fly in toward the center; nevertheless the wind becomes the factor determining the direction of swarm movement. Within the swarm, the locusts tend to space themselves a few inches apart, apparently maintaining this "interlocust distance" by sight and by the sound of the wings of their neighbors.

R. C. Rainey and his co-workers at the Anti-Locust Research Centre have made a special study of mass migrations of the desert locust. They have used aircraft to track swarms, and on occasion have followed individual swarms for more than a week. Some swarms have been estimated to contain up to 10 billion insects and to occupy hundreds of square miles. Desert locusts are said to weigh about two hundred tons per square mile and to be capable of eating their own weight in a day. Some swarms resemble stratus clouds in being well spread out but only a few dozen yards thick, while others resemble cumulus clouds in that they billow up to a great height. These different formations are related to air currents and particularly to thermal convection currents from the ground. When suitable updrafts occur, locusts may cease their flapping flight and simply glide along, thus conserving their limited fuel supplies. At night, the locusts settle on vegetation, generally resuming their flight the next morning as soon as the sun is well above the horizon. A swarm may cover one hundred to two hundred miles in a day.

Dr. Rainey and his group have gathered great masses of data on the occurrence of desert locust swarms at specific times and places, and have superimposed this information upon meteorological charts, with interesting results. It is known that the winds near the earth's surface tend to blow seasonally in certain directions, into so-called zones of convergence, where there is an excess of inflowing over outflowing air. These areas are characterized by rising air, low barometric pressure, and frequent precipitation. By careful plotting of the progress of swarms, Rainey has shown that they follow closely the progress of the "intertropical front," marking the seasonal limit of the intertropical convergence zone. Thus it is neither a coincidence nor a deliberate action on the part of the locusts that these insects commonly arrive in an area at about the same time as a storm. The locusts require moisture to complete their life cycle, and they have evolved swarming behavior that exploits weather cycles to their advantage.

That locusts arrive and depart with the wind is not exactly a new idea. According to Exodus, "the Lord brought an east wind upon the land all that day, and all that night, and when it was morning, the east wind brought the locusts." Upon the Pharaoh's repentance, "the Lord turned a mighty strong west wind, which took away the locusts." Substitute "intertropical front" for the Lord, and Exodus sounds like a modern treatise on locust behavior. Now that we have removed some of the mystery from the matter, we should be able to prepare ourselves for locust onslaughts. Unfortunately it is not a simple matter; after all, we understand weather cycles pretty well but still cannot always predict the weather accurately in specific localities. Like the weather, locusts are both predictable and unpredictable.

There is reason to believe that some locust species are declining simply because their "permanent breeding areas" have been so taken over by man that they no longer allow the great locust build-ups they once did. This may be true, for example, of the Danube delta in Europe and the Niger basin in Africa, once major breeding places of the migratory locust (and of course it may be true of our own West and the Rocky Mountain locust). At the same time, the advance of agriculture may be making things better for the desert locust and certain other species. The natural home of the desert locust is in islands of greenery in the desert, and as man has irrigated more and more of the desert he has created more and more oases suitable for breeding by this locust. Unlike some species, the desert locust has no single "permanent area," but is capable of building up in many different places and swarming over a very broad band of Africa and Asia. It is a true nomad, like many of the peoples of the desert, and in spite of all our progress in understanding him and all we have learned about insect control with fancy new chemicals supplied from airplanes, the desert locust is still one of the world's major pests. The South American locust, so long the scourge of the pampas, is related to the desert locust, and like it may prove a tough insect to conquer.

Locust research has brought out again and again the fact that each species is unique and must be studied independently—a point we have made before and shall make again. Not only does each species have its own particular color pattern and structural configuration; it also occupies its own particular home range and habitat, and responds to various factors in the environment in its own particular way. Observations and experiments on the behavior of one species are often not fully relevant to another. For example, only the desert locust males are known to produce a pheromone that hastens maturation. The migratory locust is believed to produce an inhibitory substance that actually holds back the maturation of the more advanced individuals—thus producing a similar synchrony in development, but in a different way. Similar external stimuli are known to influence differently, in these two species, the size of endocrine glands and the amount of hormone they produce—with consequent important differences in behavior. Add to this the fact that besides the various locust species there are hundreds of kinds of grasshoppers of mildly or sporadically gregarious habits (many of them showing some manifestations of phase change), and one can begin to glimpse the problems waiting to be studied.

Such is my admiration for the locust that I would gladly let him eat up my garden; it would at least be a pleasant switch from having it trampled and scratched up by the local cats, dogs, and children. But of course I don't have much of a garden, and rely on the local supermarket for sustenance. In the broad view, the locust must be regarded as a major rival of man for the earth's productivity, particularly in marginal areas that must eventually be brought to full productivity if we hope to continue our own experiment in overpopulation, as we evidently do. If we can make the other species join the Rocky Mountain locust in extinction, we shall certainly do so. At the same time we shall have lost something that we shall not find again in the farthest recesses of space. This is not to say that we should take the "anti" out of locust research, but only that we should salute the locust as one of the treasures of the universe. May his tribe decrease; but may he ever hold a place in human history and in the annals of science. And who knows, he may yet teach us something about the fundamentals of population control.

Parasitic Wasps, and How They Made Peyton Place Possible

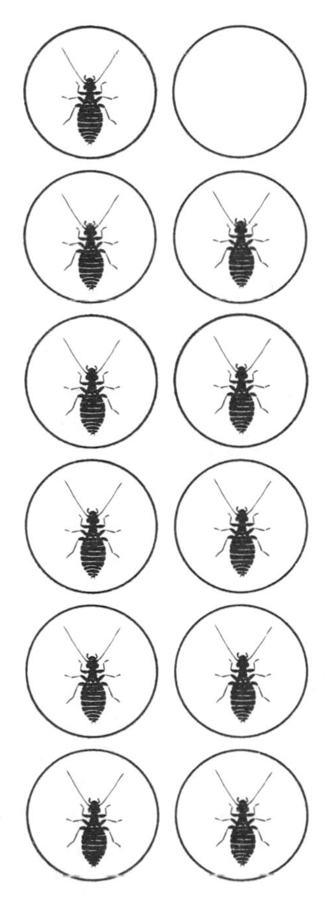

"Entomology," wrote Thoreau in the *Natural History of Massachusetts,* "extends the limits of being in a new direction, so that I walk in nature with a sense of greater space and freedom. It suggests, besides, that the universe is not rough-hewn, but perfect in its details. Nature will bear the closest inspection; she invites us to lay our eye level with the smallest leaf, and take an insect view of its plain. She has no interstices; every part is full of life."

The dramas enacted on a leaf or a stem put to shame the jungle or the African veldt—if we can but diminish ourselves to the proper scale. Aphids (plant lice) loom like fat, sluggish cattle, now and then spawning an aphid calf rather than laying eggs like respectable insects. Ants loom above them like gigantic robots, milking them for their honeydew and pointing menacing mandibles at aphid lions (the larvae of lacewing flies) lurking in the background. Predaceous fly maggots pick off aphids from the edge of the pack, and parasitic wasps hover about and now and then slip in to pierce an aphid and lay an egg inside. A neighboring leaf is occupied by a caterpillar that is devouring it in great crunching swaths, turning it into brown pellets and into the stuff of its own butterfly potentialities. But a hunting wasp swoops down, and in spite of all its thrashings the caterpillar is stung into a deep sleep, its substance suddenly fated to produce not a butterfly but a wasp. In the meantime a scavenger beetle has snapped up one of the droppings and hustled it off into a hole. The midges dance, a tree cricket shrills, and a deer mouse scratches a flea.

It was a long while before man appreciated that creatures as small as these were indeed "perfect in their details"—that each contained fully formed and discrete organs for breathing, digesting, and reproducing, that each contained a complete nervous system, and sense organs sometimes superior to our own. It is said that Galileo, in 1610, perhaps satiated for the moment with the rings of Saturn, peered through the "wrong end" of his telescope and studied the compound eye of an insect—thus with one instrument demonstrating that there are worlds both vaster and more minuscule than our own. It remained for Galileo's fellow countryman Marcello Malpighi (who was eighteen when Galileo published his major work) to describe the internal anatomy of an insect for the first time, using a greatly improved microscope. Malpighi proved that, far from being devoid of internal organs, insects are miniaturized, highly complex beings. It is ironic that Malpighi was honored by having his name permanently attached not to one of the more charming features of insects but to their excretory organs, the Malpighian tubules. Sometimes it hardly pays to be a pioneer.

One does, of course, find a certain economy in the construction of an insect (a housefly, by the way, is probably somewhat above average in size for an insect). The body wall and lining of the digestive tract are only one cell thick; there are no lungs or red blood cells; the muscles and smaller sense organs are often served by only one or two nerve cells, and in fact the entire body has only a few tens of thousands of nerve cells—which is hardly any by our standards. Insects are incapable of attaining large size because of limitations imposed by their breathing system—small tubules that carry air directly to the tissues—and by the fact that an animal with its skeleton on the outside is likely to collapse of its own weight if this exceeds a certain maximum (especially when the body wall is soft, at the time of molting). This being so, insects have evolved ways of making the most of their relatively small bodies.

All insects are specialists of one sort or another: that is, an insect does a few things extremely well, and most things not at all. Confine a mosquito with some bean leaves and it will do nothing but wither up and die; but a bean leaf is the whole universe so far as a bean beetle goes. A male promethea moth can smell the female from a mile away, if the wind is right; but that is the only odor he responds to. The behavior of some insects is marvelously complex. But put them in a strange situation, and they are utterly confused. Insects are programmed to respond to only a few elements in their environment, and to respond in only a few ways. However, so many different kinds of insects have evolved that collectively they can do almost anything not requiring them to reason or to learn very much. A given insect simply doesn't have enough nerve cells, enough plasticity of form and function, to depart very much from the specific role it has evolved to fill. That there are so many kinds of insects doing so many subtly different things is the eternal frustration of entomology—and the source of much of its fascination. Oliver Wendell Holmes entertained an entomologist in *The Poet at the Breakfast Table*, resulting in the following dialogue:

" 'I suppose you are an entomologist?'

" 'Not quite so ambitious as that, sir. I should like to put my eyes on the individual entitled to that name! . . . No man can be truly called an entomologist, sir; the subject is too vast for any single human intelligence to grasp.' "

A look at some of the very smallest of insects may help us to appreciate some of the complexities—and simplicities—in the behavior of these minute creatures. Trichogramma is a parasitic wasp that develops inside the eggs of other insects, sometimes several inside one egg. It is only a fraction of a millimeter long—so small that it would just about cover up one of the lines on an average ruler. The first account of the life history of this minute insect was provided in 1799 by a man justly regarded as the first American entomologist, William D. Peck (the insect was not even given a scientific name until some decades later). Peck, son of a Maine farmer and shipbuilder, published several pioneering studies on insects while living quietly on his father's farm and before being appointed, in 1804, as Harvard's first professor of natural history. Peck was studying the "slug worm," a pest of pears and cherries, and found that some of its eggs were not translucent but dark and opaque. From these parasitized eggs he reared a minute wasp "of a pale rust color; the eyes and three spots [ocelli] on top of the head, of a bright red. . . . I observed this year that great numbers of the eggs of the slug-fly . . . were rendered abortive by this atom of existence."

Peck provided an excellent drawing of Trichogramma, made by himself through a homemade microscope, and prefaced his paper with a fitting quotation from Pliny, the Roman naturalist-philosopher of the first century A.D.: "Natura nusquam magis quam in minimis tota est" (Nature is nowhere more perfect than in the minutest of her works).

Trichogramma has certain advantages as an experimental animal despite its size—in part because of it, since many an entomologist finds himself trying to function in a space hardly big enough for himself, and he can keep a few hundred Trichogrammas in a small vial. Furthermore, this insect requires less than two weeks to go through its life cycle, so one can rear many generations in a short time. One does need a good supply of insect eggs, but even here Trichogramma is cooperative, developing as it does on the eggs of a great variety of insects. Most researchers have found it convenient to rear them on the eggs of unnatural hosts such as flour moths, which can be kept going easily in the laboratory on flour, oatmeal, or similar substances. A variety of convenient techniques have been developed for mass-rearing flour moths and their parasites.

A number of years ago George Salt, of Cambridge University (whom we also ran into while surveying an English pasture in Chapter 2), began a series of studies on Trichogramma evanescens designed to bring out some of the details of the *modus operandi* of so small an insect. Salt was interested, first of all, in finding out some of the criteria Trichogramma uses in selecting its host. Although these wasps lay their eggs and develop successfully in the eggs of literally hundreds of species of insects belonging to several orders (including beetles, true bugs, flies, and especially moths) they do not, in fact, successfully attack *all* insects. Furthermore, if in a given area they are presented with a choice of several kinds of insect eggs, do they attack all equally or do they have

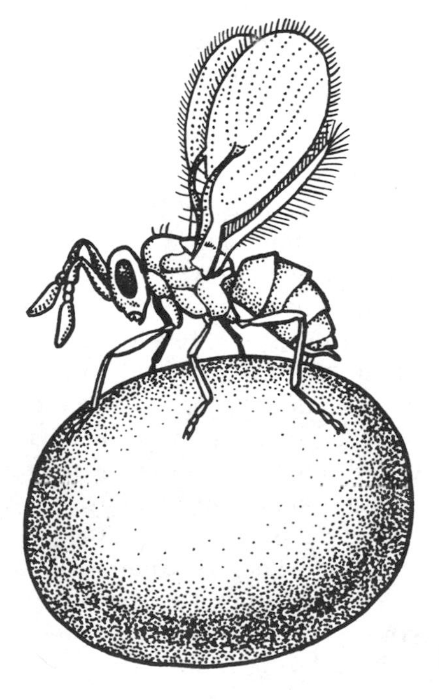

A female Trichogramma laying her egg inside the egg of another insect. This minute, yellowish wasp, with bright red eyes, is so small that it is barely visible to the naked eye.

certain preferences? Do wasps reared through successive generations on one host show a preference for that host? These are some of the questions Professor Salt asked of his cultures, realizing of course that results obtained in a bottle and on an unnatural host may have to be applied with caution to natural conditions.

Salt started out with two different stocks, one reared through many generations on the Mediterranean flour moth, the other through many generations on the so-called Angoumois grain moth (after the province of Angoumois in France, where this insect was first found to be a pest). When given a choice of the eggs of the two hosts, females of both strains showed a greater tendency to lay their eggs in those of the Mediterranean flour moth. The eggs of this moth are larger, rounder, smoother, and less brittle than those of the Angoumois grain moth—and presumably the eggs of the two species taste or smell slightly different. On what basis does Trichogramma choose the flour moth?

Salt approached this problem by using the eggs of still other insects, some of which were smooth, some rough, and so forth. In every case Trichogramma females showed a preference for the larger eggs, regardless of their other properties. They even showed a preference for the eggs of the bean weevil over the slightly smaller eggs of the flour moth, even though their offspring do not develop to maturity in the bean weevil. If size is the overriding factor, and such things as surface texture and odor are unimportant, will the wasps accept "false eggs" provided they are larger than the real eggs available? Salt observed one hundred attacks on a random assortment of grain moth eggs and sand grains selected as being slightly larger than the eggs. Surprisingly, two thirds of the attacks were on sand grains, which of course the wasps could not pierce with their ovipositor (egg-laying tube), and only one third on the moth eggs. The same results were obtained with lobelia seeds, which are slightly larger than moth eggs; but given a choice of moth eggs and sand grains selected as being smaller than the eggs, the wasps preferred the eggs.

These and other experiments enabled Professor Salt to define some of the criteria of host selection in Trichogramma. The wasps prefer larger eggs (or other objects) to smaller ones; the object must be motionless; it must be exposed (or at least not deeply buried); and it must be sufficiently firm to support the wasp, which examines it by walking on it. Some of Salt's most interesting experiments involved the use of globules of mercury as "false eggs." Although these are liquid, they have sufficient tensile strength so that the wasps can walk on them, and they readily accept them even though they cannot pierce them with their ovipositor. The size of the mercury globules could easily be varied, and Salt found that a globule many times larger than the wasp and weighing nearly half a gram was attacked; this globule had a volume nearly 2,000 times that of the smallest egg attacked. Apparently it provided a supernormal stimulus similar to those we found to occur in certain butterflies. Salt found that although shape was relatively unimportant—even square crystals of calcium carbonate were attacked—very slender glass rods were not accepted even though their volume was within the acceptable range. Apparently the object has to be of sufficient diameter to contain the progeny of the parasite.

Thus we see that this minute insect is responsive to only a very few of the qualities of the objects it encounters in nature. It is fooled completely by absurdly different sorts of objects, just as a damselfly may be duped into courting a brightly colored fishing float or a male mosquito into mating with a piece of gauze when the wing sound of the female is simulated on a tuning fork. Of course, Trichogramma does not encounter mercury globules, glass rods, lobelia seeds, or even sand grains as she flies about in vegetation searching for eggs to parasitize; everything about her is "stripped down" to just what is needed to get along in a particular situation, and no other. Actually, there is evidence that in nature the odor of the host eggs does provide an important stimulus. Working as he was with unnatural hosts, which lack the true host odor, Salt had eliminated this factor—demonstrating at once the value of simplified experimental setups and the danger of extrapolating to natural situations.

In fact Salt did find odor to be important in another facet of egg-laying behavior. When confronted with small host eggs, each capable of nurturing only one parasite larva, the female is at an advantage if she can avoid laying an egg in a host already containing an egg or larva of Trichogramma. George Salt found that when females were introduced into chambers containing some parasitized eggs and some unparasitized eggs, they almost invariably selected unparasitized eggs. However, when he washed the eggs the females were unable to discriminate between parasitized and unparasitized eggs—up to the point when the ovipositor was inserted. Then, if the egg contained a parasite, it was quickly withdrawn. Apparently the wasp detects an odor left by a female that has been on the egg previously, but if this odor is artificially removed, she is still able to detect an internal difference in the egg, probably by sensory hairs on the tip of the ovipositor. Of course, a "false egg" containing the spoor of the female is also rejected.

When a female is confined in a chamber containing nothing but eggs that are already parasitized, she is able to hold back her eggs for a time, but eventually some eggs are laid in parasitized hosts. In this situation, if she is given a choice of larger parasitized eggs and smaller parasitized eggs, she as usual selects the larger ones. In these eggs her progeny have a much greater chance of surviving—in nature, five or more Trichogrammas have been known to develop successfully in relatively large eggs. But the deposition of more eggs than the host egg has nutrient to support results in unsuccessful development: the offspring tend to be mostly males, or they tend to be dwarfed and imperfect, or in extreme cases they simply die before reaching maturity. It is as though the female could foresee this, but of course she can't. Her behavior is highly adaptive—that is, it has evolved to serve the best interests of the species—yet it is behavior of far greater simplicity than it appears at first. What we call "instinct" often turns out to be of this nature: far from inheriting some sort of mystical knowledge of what it is supposed to do, an insect simply responds in limited ways to a limited number of key stimuli. Since these responses have been molded by natural selection down through the ages, they are the ones that lead the species to fill successfully some particular, limited role in the drama of nature.

Work on a different species of Trichogramma, which occurs in nature primarily in swampy situations, led Professor Salt into some slightly different aspects of this subject. This species, Trichogramma semblidis, is a parasite of the alder fly, a dusky-winged insect that lays its eggs over water and spends its larval stage in the water. Both sexes differ very slightly (but consistently) from Trichogramma evanescens in the nature of their antennae, but the males of semblidis differ greatly in that they are always wingless. However, when Salt brought these into the laboratory and reared them on the eggs of flour moths, the males turned out to have fully developed wings. Since the eggs of the alder fly are larger than those of the flour moth, it was hard to believe that males emerging from alder fly eggs were imperfect or deformed. Evidently something about the "quality" of the host determined whether or not the males had wings. This was an especially interesting conclusion, since we know that the castes of social insects are determined at least in some cases by the quality of the food they receive as larvae. For example, a change in diet on the third day of larval life determines whether a worker or queen honeybee will be produced: the presence or absence in the diet of "royal jelly," a product of certain glands of the workers, being the deciding factor. The fact that smaller individuals often emerge from smaller hosts has been well known for a long time, and this, too, may presage conditions in the social insects, since we know that queens often receive more food in the larval stage than do workers.

That kind and amount of larval food can influence both size and major structural features of adult insects also has many implications to the taxonomist, who, in the absence of experimental work, often can not be sure whether he is working with one species or several. We now know that such things as rate of development, sex ratio, and perhaps even behavioral responses of the adults can be influenced by the kind of host on which the parasite developed. Rate of development and sex ratio may also be influenced by temperature. Some twenty-five years ago Stanley Flanders, of the University of California at Riverside, was engaged in rearing a small wasp parasitic on the red scale insect, a pest of citrus in California. When reared on mature scales developing on citrus at 80° Fahrenheit the parasites produced only female offspring, and these females were perfectly capable of laying female-producing eggs in the complete absence of males. This is not a terribly rare phenomenon in insects: quite a few parasitic wasps reproduce uniparentally, as do some beetles, and even a few cockroaches, as we saw in Chapter 3. But when Flanders reared his parasites on immature scales, or on scales developing on squash, an unnatural host, quite a few males were produced. When he raised the temperature to 90 degrees during the egg and early larval stages, some of the resulting females produced nothing but male offspring. More recently, working with a different parasite, Professor Flanders succeeded in mating such "temperature-induced" males with females that had been subjected to high temperatures in early stages of their development. In this way he developed a biparental strain of a normally uniparental species, and the biparentalism persisted for one additional generation at room temperature.

Frank Wilson, of the Commonwealth Scientific and Industrial Research Organization (CSIRO) in Canberra, Australia, has shown that the sex ratio can sometimes be altered by exposing adult female wasps to high temperatures. He was working with Ooencyrtus submetallicus, a species so small that one could put several hundred of them in the space occupied by its name. This wasp is an egg parasite of the green vegetable bug, and was introduced into Australia from the West Indies for the control of that insect. The parasite was formerly known from females only, but Dr. Wilson found that his cultures of this parasite did contain an occasional male. In the course of twenty months he reared about 15,000 females, and among them he found 126 males, as well as six other individuals that had a mixture of male and female features. The males, however, were pretty useless individuals, and seemed to lack some of the instinctive responses one expects of this sex. They pursued females and stood before them, face to face, when they came to rest, but they did nothing further. "Sexual behavior," Wilson remarks, "has evidently degenerated to a stage at which the adults do not proceed to copulation." Apparently the insects have not only carried sexuality to extreme heights, but in this instance have carried Puritanism to its ultimate.

Wilson found that this parasitic wasp reproduces best at a temperature of about 80° Fahrenheit. At 84 degrees it still does well. At 85 degrees the usual crop of females emerges, but their progeny consist largely of males and intersexes (individuals not wholly males or females), and within a very few generations at this temperature the line dies out. At 86 degrees or higher, the next generation consists entirely of males or intersexes, so the line dies out immediately. In this case it is the temperature to which the egg-laying female is exposed that is critical; the temperature at which the parasite's egg or small larva develops is unimportant. Several other cases are known in which exposure to abnormally high temperatures, either in the immature or in the adult stage, causes a shift toward the production of males, but we do not know how this effect is brought about.

Frank Wilson has also studied another egg parasite of the green vegetable bug, a wasp called Asolcus basalis, and a member of an entirely different group of parasitic wasps from the above species (and from Trichogramma). This wasp is decidedly a "horse of a different color." The sex ratio is about one male to every two or three females, but the males develop more rapidly, emerge before the females, and play an important role in the reproductive biology of the species. The eggs of the host stinkbug are laid in masses, and the female parasite has the ability (like Trichogramma) to detect the spoor of a female on an egg, and thus avoid laying an egg in a host already parasitized. Furthermore, when a female finds that all eggs in a given mass have been parasitized—that is, when she is unable to find any eggs that do not have the spoor of a female—she becomes highly aggressive and drives away any other females that may be on the egg mass. After a while the female that is "queen of the hill" also abandons the egg mass and looks for another. All this behavior tends to ensure that the parasite disperses itself well and lays one egg in each of the host eggs present in the area.

The interesting thing is that the males are also aggressive, and tend to defend the egg cluster from which they emerged against other males—and to mate with all the females emerging from this particular batch of eggs. The first male to emerge patrols the egg mass and rarely leaves it; he dashes at any male that approaches, and promptly drives away any male that emerges from the eggs in that particular cluster. When he observes a parasite chewing its way out of the host egg, he remains there and touches the emerging wasp continually. If it turns out to be a female, he mounts her immediately, and after copulation she flies away and searches for an unparasitized egg cluster. Wilson isolated one male and presented him with a series of virgin females to see how many he would mate with. Over a period of five days, the male was provided with seventy-two females, and he mated with all of them. (Advocates of stronger legislation on the treatment of laboratory animals please note: scientists do not always torture their victims!)

Thus male territoriality and aggressiveness, which we observed in the cricket in Chapter 5, crop up again in these very different insects—as they do indeed in a wide variety of unrelated animals. Yet some parasitic wasps are able to dispense with the male sex entirely. All this illustrates a point I have made many times: that every species is different, and has to be studied of and by itself before we can generalize very much about it. The parasitic wasps provide innumerable examples of this, though unfortunately only a few of them have been studied in as much detail as Salt has studied Trichogramma and as Wilson has studied Asolcus. Let me cite a few other examples of dramatic biological differences among parasitic wasps.

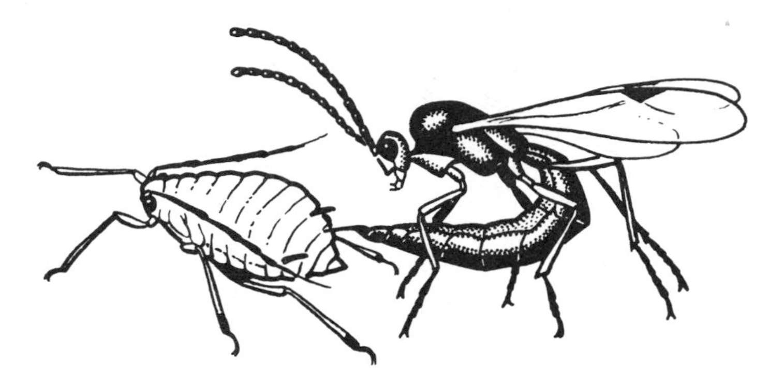

A small braconid wasp that specializes on aphids. The "wasp waist" permits the parasite to bend its abdomen beneath and forward and to lay its egg inside the aphid while the latter is in front of it. The parasite larva will develop inside the aphid, causing it to swell and turn dark; the adult wasp will cut a circular hole in the aphid when it emerges.

We saw that Trichogramma attacks the eggs of a great variety of insects, accepting almost any that occur in its habitat and are of a certain size. Many other parasitic wasps are also not at all host-specific, but take almost any insect occurring in a particular habitat. Browsing through the Catalog of Hymenoptera produced a few years ago by workers of the United States Department of Agriculture, one finds, for example, that half a page is required to list (in small print) the known hosts of one common ichneumon wasp. This is a much larger wasp than Trichogramma, and lays its eggs in the larvae of many different caterpillars; it is also able to develop as a parasite of certain wasp larvae occurring within caterpillars. That is, it is capable of developing as a hyperparasite (a parasite of a parasite). On the other hand, Asolcus basalis is a specialist on the eggs of the green vegetable bug and a few related species—and this is a rather more common condition. In one of the largest genera of parasitic wasps, Apanteles, there are several hundred North American species, each of them attacking only a rather limited number of related caterpillars. (One of the best-known species, Apanteles congregatus, attacks the tomato and tobacco hornworms, and is well known to gardeners because of the many white cocoons its larvae spin on the surface of the host.) In these cases, the parasite is guided by certain stimuli to the habitat of the host, and then (usually by odor) to a specific kind of insect.

In some cases the plant on which the host insect is feeding is of overriding significance to the parasite. For example, Apanteles congregatus develops normally on hornworms feeding upon tomatoes and most kinds of tobacco, but on certain kinds of tobacco, although the hornworm develops normally, the parasite dies before reaching maturity. In such cases it is assumed that a toxin in the plant (perhaps similar to those discussed at the conclusion of Chapter 9) inhibits the parasite but not the plant feeder. For many years the State of California tried unsuccessfully to import a certain small wasp parasitic on the red scale of citrus. The entomologists concerned always used sago palm for importing parasitized scales, since the scale develops normally on the palm and they wished to avoid shipping in citrus stock that might be infested with other pests or diseases. After thirty-five years of trial and error, it was discovered that scales grown on sago palm are immune to parasitism by this wasp, and by a simple change of tactics they were able to establish it successfully.

In cases such as this, the parasite attacks one or several insects that may occur on a variety of different hosts, and one assumes that the egg-laying female responds not so much to the kind of plant as to stimuli emanating from the host insect itself. In a number of instances it has been shown experimentally that the parasite is highly responsive to the odor of the host. The importance of odor was shown dramatically a few years ago by Harry W. Allen of the United States Department of Agriculture (now retired, but still very active). Allen was involved in the importation from Japan of parasites of the Oriental fruit moth, a very serious pest of peaches and plums. Unfortunately the fruit moth is not especially easy to rear in the laboratory, so Allen tried producing the parasites on a related moth much easier to rear, the potato tuberworm. One particular ichneumon wasp was found to develop normally in the potato tuberworm, but the females showed very little interest in laying their eggs on them. However, he hit on the scheme of spraying infested potatoes with extracts of Oriental fruit moth larvae, and that did the trick. Even better, he found that after about a dozen generations, the parasities showed signs of developing a race that would attack the potato tuberworm without having to be deceived in this way. After fifty generations he obtained a stock that bred very successfully on the new host. Thus he showed that instinctive responses are not absolutely fixed and unchangeable—as of course they cannot be if evolution is to occur.

Host recognition is ordinarily preceded by habitat-finding; that is, the wasp first has to find the place where the host is likely to occur. Sometime the parasite is guided by odor. For example, some of the parasites of blowflies are known to fly about freely until they find themselves in an area containing the odor of decaying meat; then they undergo rapid turning movements that enable them eventually to sight the pupae of the blowfly. Some of the parasites of flour moths are said to be attracted to the odor of oatmeal even when it is not infested with moths. In other cases the attraction to the habitat is apparently visual. A common parasite of the larvae of ant lions (often called "doodle bugs") first flies to sandy places and then examines depressions in the sand—where the ant lion is likely to be found.

A nice example of the attachment of a parasite to the habitat of its host is shown by a very small wasp that attacks the eggs of water beetles: Caraphractus cinctus. The adult wasps emerge beneath the water and often spend their entire lives there, swimming about with their wings. However, if a female is unable to find a host or the male a mate, the wasp may emerge from the water and skim over the surface with rapidly vibrating wings. Miss Dorothy Jackson, of St. Andrews, Scotland, who has recently been studying these wasps intensively, finds that mating may occur either beneath the water or above it, but the surface film provides "a complete barrier to recognition"—both have to be above or below it. When a wasp reenters the water, he or she may walk down a plant stem or simply dive in head-first, the front legs pointing forward and the antennae folding backward. As soon as a stem is reached, the insect pauses and cleans off the air bubbles that cling to it.

Caraphractus, like Trichogramma, is able to avoid laying her eggs in a beetle egg that is already parasitized. In this case any odor would, of course, wash off the host egg, so she must pierce it first. Evidently small sense organs on the ovipositor tell her whether or not the egg is parasitized. If the egg is small, she generally lays three eggs in it, two of which will produce females and one a male. However, when Miss Jackson made many host eggs available in quick succession, the wasps laid mostly female eggs. After egg-laying, the female walks backward and forward, tapping the beetle egg with her antennae. This apparently is her way of ascertaining the size of the host, and if it is the egg of one of the larger water beetles she continues to lay eggs rapidly in the same host. In some cases as many as sixty-five parasites have been found to emerge from a single beetle egg, most of them females. It is said that when the first male emerges from such an egg he remains there and mates with the females as they come out. In instances of this nature, the mother wasp has "made the most" of her eggs—the majority produce females, and only enough males are produced to ensure that all the females are fertilized. The male must, of course, inherit behavior patterns that ensure that he does just that. This usually means mating with his sisters, but this does not seem to have any serious effects on the wasps (there may be just enough out-crossing to mitigate the effects of continual inbreeding).

That most female wasps are able to lay either male or female-producing eggs is well known. Males are produced from unfertilized eggs, females from fertilized eggs, and sex is determined by controlling the flow of sperms from those stored in the sperm pouches of mated females. This was first discovered in the honeybee in 1845 by a Silesian priest, Johannes Dzierzon, and is known as "Dzierzon's law." Virgin females of many wasps will lay eggs, but since they have no sperms in their pouches, all the eggs produce males. An exception is provided by those species in which the male sex is absent or nonfunctional; in this case formation of the eggs is different, and all produce females—unless the mother is subjected to unusual stresses, as mentioned earlier.

This matter of "getting the most" out of the eggs is carried to an extreme in certain wasps in which the embryo divides many times to produce several offspring from a single egg. This is called "polyembryony" (literally, many embryos). Accidental polyembryony occurs in many animals—multiple births in humans, for example, are sometimes of this nature. Polyembryonic offspring are identical and of the same sex, and an animal that is able to reproduce this way regularly is able to produce a great many offspring from relatively few eggs. Parasitic wasps of several groups are known to exhibit polyembryony, producing anywhere from two to a hundred or more offspring per egg. A very small wasp called Litomastix, for example, lays its eggs inside the eggs of the cabbage looper, a very much larger insect. The cabbage-looper egg hatches, and the caterpillar feeds and grows; in the meantime the parasite egg splits into many embryos, and these produce small larvae that develop within the growing looper. Finally, the host dies when it is nearly full grown, its body packed with parasite larvae. In one case 3,000 Litomastix emerged from a single cabbage looper. In this instance it is believed that a number of eggs were laid in the host. "Mixed broods,"

A dead and distorted cabbage looper caterpillar whose body is completely filled with pupae of the polyembryonic parasitic wasp Litomastix.

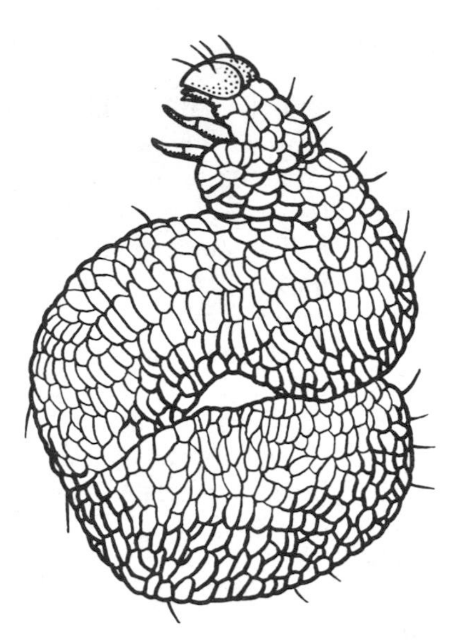

containing both sexes and the offspring of more than one female, can of course result if more than one parasite lays her eggs in the host.

Parasitic wasps exhibit so many unusual phenomena, and these are combined in so many different ways in the various species, that there seems no limit as to what to expect of these insects. Some species first sting the host; that is, they inject a paralytic venom serving to quiet the host during egg-laying. Depending upon the species of parasite and size of host, one or several eggs may be laid at a time, employing either one or several insertions of the ovipositor. The eggs may be laid internally or externally (again, depending upon the behavior of that species of parasite), and in many cases they are laid at some particular point on or in the host—a few wasps are even known to lay their eggs inside the brain or the intestine. Many parasites feed on the blood that exudes from the puncture, and some actually make a little "feeding tube" with their ovipositor, enabling them, after the ovipositor is withdrawn, to turn around and "feed through a straw" on the blood of their host. Many species require such nourishment to enable their eggs to develop, and females of such species often sting an insect and feed on the blood without actually laying any eggs at that time.

As we have seen, there are parasites that attack and develop in insect eggs, others that attack eggs but do not develop until the host eggs hatch into larvae. There are others, of course, that attack larvae directly, still others that attack pupae, and a few that attack adults. Then there are those that lay their eggs on or in a parasite already developing on or in a host (hyperparasites or secondary parasites). Cases of tertiary parasitism are also known, and even (on rare occasions) of quarternary parasitism. A quarternary parasite is a parasite of a parasite of a parasite of a parasite. If the host is a pest, then of course a primary parasite is a beneficial insect, a secondary parasite is a pest, a tertiary parasite beneficial, and so forth; but if the host happens to be a beneficial insect (for example, a caterpillar that feeds on a weed), then a primary parasite is a pest, a secondary parasite beneficial. . . . The difficulty is that a given parasite is sometimes capable of changing its role, developing now as a primary, now as a secondary parasite. Exactly this problem arose in connection with the importation and release of parasites of the gypsy moth, a program that (like many similar ones) was undertaken without a full understanding of the biology of the insects involved.

An excellent example of this nature was presented by LaMont C. Cole, of Cornell University, when reviewing Rachel Carson's *Silent Spring* for *Scientific American.* Cole's case was drawn from the work of F. L. Marsh on the natural enemies of the cecropia moth, and is reproduced in simplified form here. The cecropia caterpillar feeds on willows and other trees and is a pest of very minor proportions, but situations as complex as this (or more so) often involve major pest species. If one counts the cecropia as a pest, then Gambrus and Winthemia are beneficial; but how does one rank Dibrachys, which is capable of attacking either a primary or a secondary parasite? And what of Pleurotropis, which is harmful or beneficial depending on how one ranks Dibrachys? When we consider that each of these insects is acted upon by various predators in various ways, reacts differently to temperature, rainfall, and other climatic factors, and differs in susceptibility to various insecticides, one can begin to grasp the complexity of the situation.

One of the most remarkable instances of hyperparasitism involves the genus Coccophagus. Importation of a certain species of this genus into California for control of the black scale, an insect that sucks the sap from citrus trees, led to very peculiar results. Shipments of infected black scale hosts yielded nothing but female parasites, although in these wasps both sexes are required for reproduction, and the cultures from repeated importations therefore died out. At the same time parasitized scale insects of another species were being imported from the same region, and some of these did yield Coccophagus males as well as parasites of another genus. Was it possible that the males developed only as secondary parasites but the females only as primary parasites? Virgin females from other shipments were placed with scale insects already parasitized, and lo and behold the resulting offspring were all males!

Stanley Flanders turned his attention to this remarkable situation and showed that indeed the newly emerged, unmated Coccophagus lays her

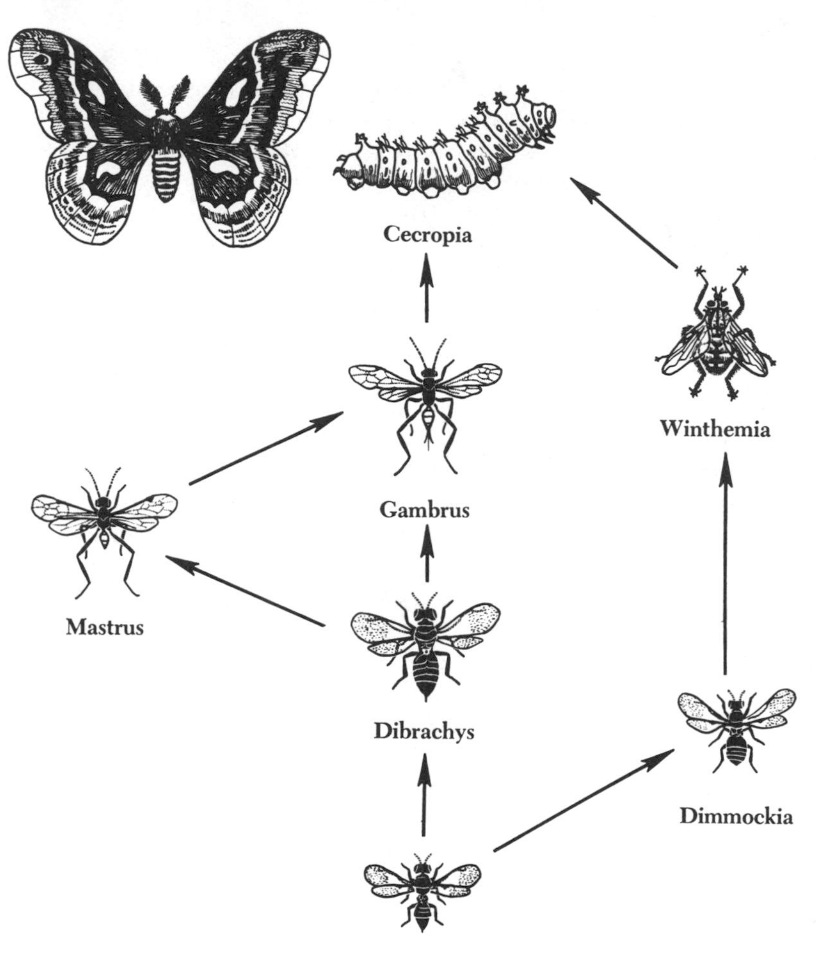

Some of the parasites and hyperparasites of the cecropia caterpillar. As if this weren't complicated enough, the cecropia has other parasites than these, and the ichneumon parasite Gambrus is attacked by at least three other wasps besides the two depicted here. Winthemia is not a wasp, but a tachina fly. (Adapted from review by L. C. Cole of Rachel Carson, *Silent Spring*.)

unfertilized, male-producing eggs only in scale insects already parasitized. But once she has mated, she "changes her instincts" and selects unparasitized scales for her fertilized, female-producing eggs. In some forty species of Coccophagus and related genera the male develops as a parasite of the female, often the female of his own species! There is actually a good deal more to the story than this, since each of the species of Coccophagus may be markedly different in its behavior. In one species the male is actually a primary parasite of scales, but in this case the male-producing eggs are laid externally on the host, and larval development is external, while female-producing eggs are laid internally and develop internally. In this species one can judge the sex of the egg while it is being laid, since the egg-laying posture of the female is different. In other species the unmated female lays her eggs more or less indiscriminately in various scale insects, but after mating she selects one particular host species for attack.

Differential egg-laying and development are carried still further by certain species of a genus related to Coccophagus in which the unmated female lays her eggs in those of certain moths, while a mated female turns to her more normal hosts, scale insects and their relatives. This unique situation came to light several years ago in Peru, where a certain parasitic wasp was being studied for the control of a caterpillar that feeds on cotton. Although the parasite was common enough in cotton fields, only males were reared from the eggs of the moths. Oscar Beingolea, of the Peruvian Ministry of Agriculture, showed that the males developed as primary parasites of the moth eggs, while the females of this same species developed as primary parasites of white flies (sucking insects related to scales) also attacking cotton. In the United States, species are known in which the males develop in the eggs of codling moths, the females in scale insects. But there is no need to continue this parade of the almost unbelievably strange reproductive strategies of Coccophagus and its relatives. Enough has been said to indicate that one cannot hope to use parasitic wasps in biological control of pest insects without a fairly profound knowledge of them—and what one learns about one species is usually not true of another species, at least not in every detail.

I have always been especially intrigued by the fairly numerous parasites that reach their host by some devious means. For example, some of the egg parasites of locusts "hitch a ride" on the female locust, then slip off when she lays her eggs. Certain egg parasites of preying mantids also do this, and in this case such behavior is especially important, since mantid eggs are covered with a frothy substance that hardens soon after it is formed. However, by being there when the eggs are laid, the parasite is able to attack the eggs before the covering has hardened. It is said that the wasps attach themselves to the mantids soon after their own emergence, and often have to remain there several months before the mantid lays her eggs. In order to survive, the wasps nibble at the mantid from time to time and lap up a little of her blood.

The females of one peculiar group of rather rare parasitic wasps, the Trigonalidae, lay their eggs not on the host at all, but on leaves. Here the eggs remain alive, sometimes for months, but do not hatch unless the leaf is eaten by a caterpillar. Once inside the caterpillar, the abrasion and the digestive juices cause the egg to hatch. Experimentally, the eggs of the parasite can be made to hatch by piercing them gently and then covering them with weak caustic potash. Curiously, the parasite does not attack the caterpillar, but one of its ichneumon wasp parasites (that is, it is a hyperparasite). Needless to say, a great many of the eggs are never eaten by a caterpillar, and many of those that are end up in an unparasitized caterpillar, so they fail to develop. The female wasp compensates for this by laying a prodigious number of eggs. One of them is recorded as laying 4,376 eggs in one day.

An even more devious method of reaching the host, coupled with a high egg-laying capacity, is shown by quite a different group of wasps, the family Eucharitidae. These wasps attack the larvae of ants, but the females lay their eggs in situations remote from ant colonies, chiefly on leaves or in buds or seed pods. Here the eggs hatch into small larvae that are more or less "clothed in armor," that is, covered with hard, spiny plates that enable them to withstand drying and to crawl about to some extent. In some cases the larvae assume a "waiting posture" on the

A newly hatched larva of a eucharitid wasp in waiting posture. If an ant comes in contact with the larva, the latter will attach itself and be carried to the nest, where it will fasten itself to an ant larva, wait until the latter transforms to a pupa, then molt to a more grublike body form and consume the ant pupa.

plant until a worker ant comes along, whereupon they grab hold of it and are carried to the nest. Others attack ants that do not forage on plants, and in this case they propel themselves into space, to fall to the ground and assume a waiting posture there. Once in the ant nest, the parasite larva transfers to an ant larva, waits until the latter transforms to a pupa, and then proceeds to devour it. The parasites emerge as adults inside the ant nest, and it is said that in some cases the ants carry them around and even feed them, though eventually they leave the nest. Mating may occur outside the nest entrance or sometimes inside. The late Dr. William Mann, for many years director of the National Zoo in Washington, once observed males mating with female pupae in the nest—before they had even transformed into adults.

All this surely comes under the heading of "useless information": these are tiny, odd-looking wasps, never noticed by the average citizen; they are only sporadically common, and all attack ants, most of which themselves live relatively obscure lives. Surely the lives of these insects are of interest only to a few professors, who are or ought to be teaching courses in insect control to justify their salaries. These are the days of cost-benefit analysis; the days when we are being urged to channel our efforts into the fields most likely to produce results to fill immediate human requirements. In President Johnson's words, the "time has now come to zero in on the targets." He was speaking of medical research, but the philosophy is now pervasive. "Urgent support of a field," said the director of the Oak Ridge National Laboratory recently, "is justified only if that field is likely in some way to solve a pressing human need."

Unfortunately, this distinguished gentleman failed to spell out his formula for deciding what "is likely to solve a pressing human need." The lowly Eucharitidae, the most insignificant of the insignificant, in fact created a bit of a stir a few years ago. Modern man has been conditioned to expect his fruit to be perfect in every detail: large, immaculate, delectable; and the fruit industries are geared to the production of such fruit. The modern banana is a jewel in the crown of agricultural technology. But a few years ago the United Fruit Company found some peculiar spotting on the skin of its bananas. The spotted bananas were not wormy, but they could not be sold for a good price. Observation revealed that the spots were caused by a strange-looking little insect having no known connection with bananas—a eucharitid wasp. No one knew much about them, but by digging up the researches of a few starry-eyed naturalists it was concluded that this must be an ant parasite that was merely using the bananas as a place to deposit its eggs, such that the larvae would have a good "jumping-off" place prior to attaching themselves to an ant on the ground. Suddenly these poorly known insects were found to be affecting the Gross National Product! This was, of course, an unbearable situation; nothing is more sacred than the GNP.

Parasitic wasps are not often cast in the role of pests. More often, in these post–*Silent Spring* days, we tend to think of them as diminutive knights in armor, brandishing their ovipositors in the morning sun and about to save man from the horrors of his own technology—that is, from the side effects of his insecticides. Rachel Carson herself had some reservations on this matter. She said:

"The predator and the preyed upon exist not alone, but as part of a vast web of life, all of which needs to be taken into account. Perhaps the opportunities for the more conventional types of biological control are greatest in the forests. The farmlands of modern agriculture are highly artificial, unlike anything nature ever conceived."

But as our population continues to double every few generations, more and more of the earth will inevitably be converted into massive farms. Are the parasitic wasps likely to play a significant role in insect control on our crowded planet? The answer is *no*—at least so long as we are preoccupied with cost-benefit analysis, with research planned so as to produce immediate and foreseeable results. The fact is that so few people have been concerned with the study of parasitic wasps, so little time and money invested in this field, that it is astounding that any success at all has been achieved by the use of these insects in the control of pest species. As I have indicated, each species constitutes a study in itself. And how many species are there? No one knows. In many groups, fewer than half the species appear to be known to science, even from museum specimens. G. J. Kerrich, a specialist on ichneumon wasps at the British Museum in London, remarks that about 50,000 species of parasitic wasps have been described, but that when and if the world fauna is fully known the number may be closer to 500,000. He thinks it possible that there may be a million species of ichneumon wasps alone, and they comprise only one of several families of this group. It is generally agreed that entomophagous (insect-eating) species of insects outnumber the plant-feeding species, perhaps by as much as two to one or three to one.

As mentioned in Chapter 1, I am something of a specialist on the family Bethylidae, the larvae of which attack various small moths and beetles. To be a specialist on these wasps is really not much of a distinction; only one other person in the world is actively working on the group. Yet there are several thousand species of bethylid wasps—we do not even know them well enough to make an intelligent estimate of the number there may be.

And, as I have said before, once one names a species and puts a label on his museum specimens, what does he know about them? If we are to make an effort to understand their role in nature, we must know something about their behavior: what and how many species they attack, in what stage they attack it, how they locate and recognize it, how many eggs they lay and where they lay them, what the sex ratio is, how their mating behavior influences their distribution and fecundity, and so forth and so on. But that is still only a beginning. How are they influenced by the host? Do they have hyperparasites? Do they compete with other parasites for the host? How do temperature and rainfall affect them, and how do these factors influence the host and the other members of the community? How do all these factors interact to produce population expansions and depressions? Are there ways of rearing a parasite and releasing it so as to cause a reduction in the host population sufficient to please the farmer and the consumer? Can biological control be used concurrently or alternately with control by insecticides?

Much has been written on all these topics, but we are still "babes in the woods" because of the overwhelming shortcomings at the very bottom of the pyramid of knowledge: the level at which "cost-benefit analysis" becomes a nonsense phrase. We have to know, as a very first step, what the species are that we are dealing with. A few years ago attempts to control the California red scale on citrus trees in that state by the use of parasites of the genus Aphytis led to confusing and sometimes unsatisfactory results. Paul DeBach, of the University of California at Riverside, delved into the taxonomy of this genus and found that what was once thought to be a single species of Aphytis included, in fact, at least seven species having different biological adaptations. As a result of these studies, it has been possible since 1957 to introduce several promising new red scale parasites into California. Differentiation of these species was first made on the basis of differences in behavior and life histories, and only later were minor structural differences discovered. Here was another example of "hidden species" similar to those we found to occur in crickets and in fireflies.

Considering the superficiality of our knowledge of most parasitic insects, it is surprising that there are some fairly impressive records of successful control by the release of wasps. The brown-tail moth, whose caterpillars have seriously irritating hairs and which was once a major pest of shade and fruit trees in New England, has been of minor importance ever since several parasites were introduced from its native home in Europe more than fifty years ago. The Eucalyptus snout beetle has been controlled very effectively by an egg parasite in South Africa, New Zealand, and other areas. In Hawaii, the sugarcane leafhopper and the avocado mealybug have been controlled by biological agents. According to Paul DeBach, one of the introduced wasp parasites of the black scale has resulted in a yearly saving of well over a million dollars, with a total saving to the California citrus industry of about 32 million dollars over an eighteen-year period. Several other serious pests of citrus fruit have been brought under control by parasites in California, Israel, and other parts of the world. It is noteworthy that many of the successes have occurred under the artificial conditions of intensive agriculture.

There have also been many failures. Our little friend Trichogramma was once and still is being reared all over the world for the control of a great variety of pests. The "fad" has had its ups and downs, and several times Trichogramma has failed to live up to expectations. After releasing 15,000 to 45,000 Trichogrammas per acre in Louisiana cane fields, workers of the United States Department of Agriculture concluded that the wasps were "of no value in the control of the sugarcane borer." Up to 1956 about a hundred imported parasites (not all of them wasps) had become established in the United States; however, nearly 400 others had been imported but had failed to establish themselves. Some parasites control their host to the extent of something like 10 to 50 per cent—which is hardly enough for a fruit farmer who knows that only perfect fruit commands good prices.

Among the advantages of biological control are its permanence and the fact that pests do not develop resistance to parasites the way they often do to insecticides. But a few years ago the larch sawfly, which was supposedly under control by an imported ichneumon wasp, began to become a pest again in parts of Canada. Dissections showed that more than 90 per cent of the sawfly larvae attacked were able to destroy the parasite's egg by forming a capsule around it. This was first discovered in the areas where the parasite was first introduced, suggesting that in these areas the host had had time to develop a strain immune to the parasite.

It is perfectly true that many entomologists are strongly biased against such partial and undependable measures, that, in Rachel Carson's words, "their professional prestige, sometimes their very jobs depend on the perpetuation of chemical methods." But can you blame them? They are expected to produce results. All the excuses in the world regarding our ignorance of the biology of parasites will do nothing to produce perfect apples. The manipulation of a complex, unique system such as a parasitic wasp (or any other organism) to suit our own ends requires a profound grasp of what we are about. And as we ponder these things, the world production of wheat, corn, rice, and other staples increases by the minute—after all, hundreds of babies are born into the world every minute. The entomologist cannot be blamed for creating the philosophical milieu in which he operates; he is simply trying to satisfy the farmer and taxpayer as best he can with the resources he has.

One notes a certain amount of dissension even among the practitioners of biological control. W. R. Thompson, one of the "grand old men" of this field, remarked some years ago that "the idea of biological control has now become fashionable and is tending to degenerate into a kind of superstition or fad." Frank Wilson has commented that "biological control tends to have strong supporters and vehement detractors, and to go through alternate phases of popularity and denigration." T. H. C. Taylor, who was involved in some very successful biological control work in Fiji, remarks that he knows this to be "the best of all methods of controlling pests when it works," but it "seldom works," and furthermore "the present tendency to organize the moving of parasites and predators about the world on an ever-increasing scale, despite decreasing results, is unsound and is, therefore, to be regretted."

Needless to say, most devotees of biological control would not agree with this. However, a rather strange rebuttal was forthcoming from C. P. Clausen, formerly chief entomologist of the Division of Foreign Parasite Introduction of the United States Department of Agriculture, and one of the most listened-to men in this field. Taylor (like many others) had pointed out the need for intensive work on the relationships of parasites, hosts, and their environment. Clausen remarked: "At the present time, it is difficult to visualize the manner in which these detailed ecological studies can contribute substantially to the practices of applied biological control." If this is a true reflection of sentiment in this field, then biological control is indeed bankrupt.

Personally, I would rather believe F. J. Simmonds, of the Commonwealth Institute of Biological Control in Ottawa, Canada, who discussed the frequent failures in this field as follows:

"In many instances only comparatively superficial studies have been made and the most obvious and apparently promising natural enemies have been selected as biological control agents. . . . Increased trade, and particularly facility and speed of air transport, have more than offset the effects of more stringent quarantine measures. . . . The spread of pest species will continue and afford additional opportunity for the use of biological control methods. . . . The approach . . . is suggested in which an insect pest or its complex of natural enemies should not be considered an isolated entity but in relation to the whole environment. . . . In the past a number of biological control projects have been abandoned as failures after the unsuccessful introductions of several exotic natural enemies of the pest in question. . . . Probably, in some of these, more detailed studies of these pests in their area of origin might provide the answer to successful biological control."

I find it difficult to predict the future of this field. *Silent Spring* has produced a good deal of talk about the greater use of parasites, predators, and disease organisms, as well as about the use of insects' pheromones and hormones against them: The Other Road, as Miss Carson called it. But I have not noticed any revolution. I still find it necessary to discourage many students wishing to study wasps or to go into other areas of basic entomology. There are simply not that many positions. The United States Department of Agriculture, for example, employs three taxonomists specializing on parasitic wasps, and they are so overloaded with making identifications (or trying to) that they spend little of their time on research. If I train a student in this field, I can be fairly sure he will have to earn his living teaching or selling insecticides. There remain vast unexplored areas in wasp taxonomy, and it should (in my biased opinion) be worth the taxpayer's dollars to send out a few exploring parties. I know of one group, the family Dryinidae, for which there is not at present a single authority in the world. Yet these insects attack leafhoppers, a group of no inconsiderable importance, and they do so in a remarkable manner: the females have a traplike device on their front legs, and they reach out and snap up the leafhopper, hold it in the air, sting it, and then lay their egg in it. The Catalog of Hymenoptera of America north of Mexico lists 118 species, and adds that "most . . . species appear to be still undescribed." A nation that devotes countless millions to the escape of boredom via television apparently cannot afford to inquire into these fascinating realms. (Tonight I could be watching "Peyton Place," "Occasional Wife," or "Petticoat Junction"—which is why I am writing a book instead of watching television.)

Yet the fact of the matter is, if it were not for the parasitic wasps, man and all the glorious manifestations of his culture would not exist. There would be no Peyton Place, no jukeboxes, no hermetically sealed caskets, no miniskirts, no pop art or electronic music. The vast majority of the tens of thousands of species of parasitic wasps attack plant-feeding insects, and it would be a vastly different world if these parasites did not exist. In the words of Stanley Flanders:

"It is quite unlikely that man could successfully compete with the hordes of other plant-feeding species constantly oscillating under feast or famine conditions in a world lacking natural enemy regulation of plant feeders. However, the parasitic Hymenoptera, in conjunction with other parasitic insects, were on the job regulating the densities of plant-feeding insects long before man and the other mammals evolved. Quite possibly it was the conservation of plant life thus ensured at fairly steady densities which permitted the evolution of mammalian forms of life."

It has often been noted that the rise of the major groups of plant-feeding insects—especially the caterpillars—occurred at about the same time as the fall of the dinosaurs. Did the dinosaurs become extinct because the world was swamped with insects that ate them out of house and home? Was the world then "saved for the mammals" by the rise of the parasitic wasps? We do not know, but the fossil record is not inconsistent with this hypothesis. At the very least it seems reasonable to believe that without parasitic wasps and other insects that keep leaf feeders at moderate levels, the course of evolution might have been very different: whole groups of plants might have become extinct, other, poisonous plants might have flourished, and the vast hordes of herbivorous mammals and their predators might have never evolved. In a sense we owe the miracle of humanity to the wasps.

But now that we have taken the world into our own hands we seem to have little use for the wasps. At least let us salute them in passing for what they have done for us. And let us continue for a few years to support a few hard-working wasp specialists. I would hate to have to support my family on the income from my books!

Of Springs, Silent and Otherwise

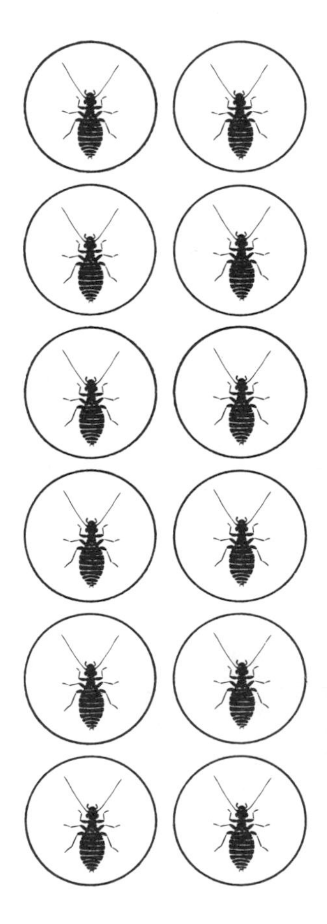

When writing a book about insects, the most difficult problem is to know when to stop. Each species has made its own particular contract with nature; each has evolved its own unique combination of structures and behavior patterns that enables it to flourish in some small segment of the planet that engendered it. Each might form the subject of a chapter of its own, if we knew enough about it—after all, several whole books have been written about the honeybee. Our ignorance of the insects is such that perfectly respectable, thoughtful estimates of the total number of species range all the way from 600,000 to 3,000,000. Professor Ross Arnett, of Purdue University, has recently totaled the number of catalogued beetles and found 219,409 species. He estimates the final total to be near 290,000. The beetles are undoubtedly one of the best-studied groups of insects, and presently they include the largest number of catalogued species. When all the parasitic wasps have been described, the Hymenoptera may well outrank the Coleoptera in size, but personally I can see little justification for an estimate of a total of much more than a million or a million and a half species of insects on earth.

That, of course, is enough—far too many, from a human point of view. Is there any point in trying to acquaint ourselves with all of them? There are biologists who claim that total knowledge of life on earth is impossible and unnecessary, that study in depth of certain representative species is enough to teach us all we need to know about the principles of life. These men are right that we should study representative animals in depth: think, for example, of all we have learned from Drosophila, from Rhodnius, from the guinea pig and the rat. Does it matter that there is a midge dancing over a patch of moss on Baffin Island, a springtail gluing its stalked semen droplets to a moldering leaf in New Guinea, a wasp living as a tertiary parasite of a miner in pandanus leaves? I would be hard pressed to prove that any of these illustrate some new biological principle or are going to impinge, at some remote date, on man's culture. Is there a need to catalogue every obscure creature on earth and to try to decipher its contract with nature? As a curator in a natural history museum, one whose job it is to preserve representatives of all forms of life, I should like to think that there is. But the view from my ivory tower is a limited one: it does not include, for example, a vista of unbroken wheat fields, the New York Stock Exchange, or the slums of Lima, Peru.

Of course, no one in his right mind would suggest dumping entomology *in toto* into the wastebasket. No one wishes to go back to 1793, when yellow fever decimated the population of Philadelphia; or to 1870, when the Colorado potato beetle seemed ready to eliminate the potato from the American diet; or even to 1940, when gardens in the northeastern United States seemed able to produce nothing but Japanese beetles. We cannot let our guard down, particularly when aircraft are capable of carrying people, produce, and pests all over the world in a few hours. Who could have supposed that the Japanese beetle, a relatively obscure insect in its native home, would suddenly emerge as a major pest on a new continent? Who would have supposed that the eucharitid wasp mentioned in Chapter 11, living as a poorly known ant parasite in Central American forests, would suddenly become a cause for concern? Who can foresee when some new insect will arrive on our shores and explode into a major dilemma—as the notorious face fly recently has? Who, incidentally, was responsible for identifying each of these new pests, thus permitting us to look it up and find out where it came from and what was known about it? In each case, it was a taxonomist, one concerned with the little-appreciated art of classifying and preserving specimens, one moved primarily by curiosity about the earth's inhabitants and little concerned with what may or may not be a "pest."

Incidentally, have you ever tried to define the word "pest"? It is derived from the Latin word *pestis,* meaning a plague, and refers to any organism that sometimes occurs in numbers inconvenient or troublesome to man. But how inconvenient and how often? A lady beetle may be a nuisance if it insists on crawling on one's leg while he is trying to take a nap in the sun; a honeybee can be far deadlier than a malaria mosquito if it frightens the driver of a car full of children—as has happened more than once. How is one to tell friend from foe in complex webs of interrelationships such as those centering upon many leaf-feeding insects, as I described in the preceding chapter? Or when the use of an insecticide will turn an "unimportant" insect into an important one, as the use of DDT against the codling moth of apples turned the red-banded leaf roller into a major pest in some areas? Who could have foreseen that modern agriculture would reduce the incidence of some locusts, but improve conditions for certain others? The idea that insects can be classified as good or bad is simply not tenable; nature is by no means that simple, much as we would like it to be.

So complex is the web of nature that few insects—indeed few plants or animals of any kind—can be categorically labeled as undesirable. Even assuming that a given species is a major consumer of an important crop, can we be sure it is worth controlling, both in terms of dollars and in terms of insecticidal pollution of the environment? What price for an apple perfect in every detail, a back yard wholly free of mosquitoes? I admit to a bias toward insects dating from my youth, for my father was not one to let me go fishing when the beans needed dusting, and I found myself sympathetically allied with the bean beetles against the latest advice from the county agent. That doesn't mean that I am against eliminating malaria mosquitoes or controlling the cornborer; I am as much in favor of alleviating hunger and sickness as anyone else. But I do think we are much too ready to tack the label "pest" on almost any insect that comes along. More than once I have encountered persons suffering from entomophobia: that is, their fear of insects was so great that they imagined they were being bitten by fleas or one thing or another. Perhaps such persons had a father who forbade them to swat flies.

Since I am an admitted admirer of the insect world, you might suppose me also to be an admirer of Rachel Carson's attack on the field of chemical insect control, *Silent Spring.* I am indeed impressed by the fact that one small voice can, in these busy times, so shake the world. The publication of her book in 1962 released a barrage of discussion, pro and con, and the issue is still far from dead. The matter has been discussed at length by the President's Science Advisory Committee, by congressional committees, in scientific meetings of many kinds, and of course in the halls of Academia. Justice William O. Douglas was perhaps somewhat carried away when he called it "the most important book of the century," but there is no doubt that few events in recent years have had a more profound impact upon the field of entomology.

Was Rachel Carson right? Are we in fact so endangered by insecticides that "we are in little better position than guests of the Borgias"? Is it true that "one in every four" of us will die of cancer, like as not as a result of ingesting carcinogenic chemicals used to control insects? Is the destruction of wildlife so widespread that our springs are, in fact, silent of birdsong? Of course, all this is nonsense. *Silent Spring* was hardly off the press before Edwin Diamond, an editor of *Newsweek* who had assisted Miss Carson in the initial work on her book, pointed out in *The Saturday Evening Post* that she had indulged in many half-truths and appeals to the emotions. In 1963, a report by the National Academy of Sciences called for more research on several aspects of the problem, commenting that "pest control to protect human health, food, fiber, forest and other biological resources is essential by whatever means necessary." In 1966, after three years of hearings, a Senate committee chaired by Senator Abraham Ribicoff of Connecticut concluded that "no significant human health hazard exists today when the great benefits of disease control and food production are weighed against known hazards." DDT, the committee reported, "has been fed to human volunteers in significant doses for 18 months without demonstrable acute effects."

It would be a mistake to forget that DDT stopped a typhus epidemic in Naples during World War II; that it has brought about a vast improvement in the health and well-being of peoples in tropical and subtropical areas through control of malaria mosquitoes. According to one estimate, DDT has resulted in the saving of 5 million lives and the prevention of 100 million illnesses. These are a good deal more impressive figures than those on the number of deaths from the misuse of pesticides: about 150 persons per year in the United States. Each year in this country, by the way, more than 200 die from overdoses of aspirin (mostly children). In a Binghamton hospital in 1962 seven babies died when salt was accidentally added to their formula in place of sugar. We live surrounded by substances, both natural and synthetic, that when consumed in large enough quantities will cause sickness or death. Carelessness is never a virtue; and insecticides must always be applied with close attention to the directions supplied with them. Recent loss of lives in Mexico and Colombia resulting from poisoning by the insecticide parathion mixed with flour should serve as a warning. It is said that as much as half of the foods we purchase contain traces of DDT, but a recent study by the Food and Drug Administration indicates that the residue of DDT in a typical diet is only one twentieth of the level rated as "acceptable" by an international committee of health experts. The FDA, by the way, is constantly on the alert for substances that may cause cancer, and when a pesticide is shown to have carcinogenic effects, no trace at all is tolerated in foodstuffs. Despite Rachel Carson's allegations, DDT and other insecticides have not been shown to be carcinogenic.

What about the effects of insecticides upon wildlife? There is good evidence that heavy dosages applied from planes or ground equipment have sometimes caused the death of fish, birds, small mammals, and a great many invertebrates other than the "target" organism, the insect they were designed to kill. The facts cannot be denied. Hopefully, we

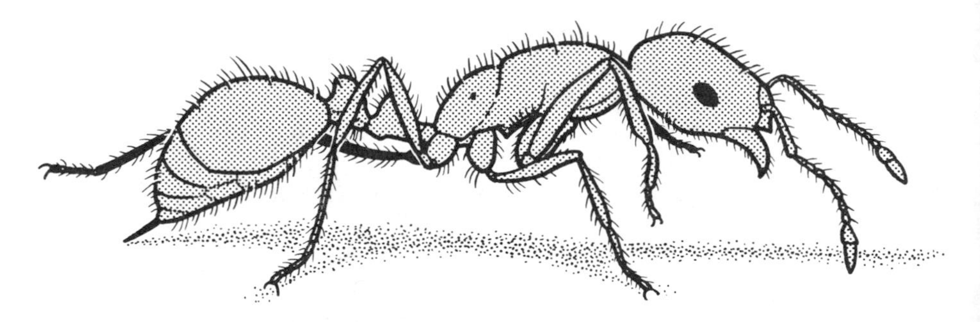

The notorious fire ant, object of an expensive and largely futile eradication campaign in the southeastern United States.

have learned a few lessons from badly mismanaged control campaigns such as that directed at the fire ant. It is said that the United States Department of Agriculture, during one twelve month period, spent more than $2,000,000 on control measures against the fire ant; but there is no evidence that, at any time, the Department spent much of anything on basic research on the behavior and environmental relationships of that insect. Is it any wonder that the campaign was a failure? This is an example of how "zeroing in on a target" without adequate knowledge of the situation can backfire; it is like trying to hit a bull's-eye with an atom bomb.

There is evidence that wildlife "kills" during our period of overconfidence in the use of pesticides were in onsiderable part both localized and temporary. So long as habitats are not actually destroyed, plants and animals (certain kinds, at least) tend to reproduce so as to refill empty niches. For example, the robin, which at one point Miss Carson carelessly refers to as being "on the verge of extinction," is obviously still very much with us. In fact, there is some evidence that robins, redwinged blackbirds, and certain other birds are more plentiful now than they were a few years ago. With regard to forest inhabitants, it should be remembered that repeated heavy defoliation and extensive death of trees (for example, as a result of the work of the gypsy moth or the beetles that carry Dutch elm disease) can cause extensive changes in the flora and fauna. In any event, the changes brought about by the use of insecticides in forested areas are hardly comparable to those caused by fires—to say nothing of bulldozers. However, I doubt if there are many cases in which the advantages of using pesticides in forests outweigh the disadvantages.

In my neighborhood, Rachel Carson's expression "silent spring" seems to have backfired. Our springs *are* becoming silent. It happens that two of my close neighbors have cats, and each spring the cats systematically pick off the baby birds one by one as they descend from their nests. To the best of my knowledge, no birds have been reared successfully for several hundred yards around my home for several years. Rachel Carson makes a great deal of the susceptibility of cats to DDT, but makes no mention of the fact that domestic cats are a major source of mortality to fledgling birds as well as to chipmunks and other small mammals. On certain oceanic islands, cats introduced by man have devastated the native wildlife. I regret that Rachel Carson allowed her sentimental attachment to cats to blind her to these basic facts of nature. One cannot have both cats and birds, just as he may have to choose, at times, between taking a chance on killing an occasional robin or letting his elm trees become infested with the bark beetles that carry the deadly Dutch elm disease.

On the agricultural scene, the bald fact is that we could not feed ourselves and the rest of the world adequately without the use of insecticides. Had it not been possible to increase by several times the per-acre yields of many crops in recent times, we would have long since been hungry and would have to pay a high percentage of our income for a poor diet. The food surpluses mentioned in *Silent Spring* are virtually all gone, and we have supplied wheat to Russia and other countries; indeed, millions of people in India subsist largely on shipments of American grain. Well over half the people of the world do not receive enough to eat. For the next few years, most increased food production will come from making land already under cultivation more productive, especially in the less developed countries, many of which realize exceedingly small yields from their soils. To talk about the "balance of nature" on the great, mechanized, one-crop farms of today is unrealistic—man has established his own balances in such areas, and he maintains them by such artificial means as cultivation, irrigation, and the use of chemical weed killers, fertilizers, and insecticides. In spite of all our efforts, each year in the United States insects account for crop losses of more than $4 billion, animal losses near $1 billion, and destruction of 8 billion feet of timber.

One of the advantages of DDT and related compounds is their "residual effect": that is, they leave a trace of poison that kills insects touching it many days or even weeks later. This was the joy of using DDT for malaria control: walls that were sprayed continued to kill mosquitoes for months. But this proved to be a mixed blessing as these toxins began to build up in soils and in plant and animal tissues faster than they decomposed. DDT has been found in the flesh of fishes in areas remote from any insecticide applications, and even in antarctic penguins and seals. Apparently contamination of the earth with these persistent chemicals is as universal as contamination with some of the products of nuclear testing, such as strontium 90 and cesium 137. We do, of course, have some very sensitive tests for the presence of minute amounts of these substances, but it is something to think about. As man remolds the world to his desires, he is repeatedly confronted with difficult choices, with weighing benefits against risks. When he uses highly potent weapons, a bad choice may be a fatal one. Supposing, for example, we inadvertently contaminated all our soils with substances that prevented the growth of nitrogen-fixing bacteria and the bacteria of decay. It would take no more than that to render the earth sterile of all life, including our own. It is already evident that runoff of nitrates and phosphates from fertilized land—among other things—is grossly altering life in our streams and lakes, and may eventually sterilize them.

There are several studies that indicate that traces of DDT taken up by the small organisms at the bottom of the "pyramid of numbers" may be concentrated in considerably greater amounts by the larger organisms that feed on them, then concentrated still further by the predators and scavengers toward the top of the pyramid. Such "biological magnification" may result in concentrations 10 to 1,000 times that at the lower end of the food chain. Recent studies of a salt marsh on Long Island, for example, showed DDT residues averaging 0.04 parts per million in microorganisms, 0.16 to 0.26 parts per million in animals the size of shrimps, crickets and minnows, and up to 75 parts per million in some of the larger birds (well within the range known to cause death). Actually, many of the larger predators probably receive concentrations not quite enough to kill them but sufficient to reduce their ability to reproduce.

Laboratory studies show that sublethal doses of DDT reduce the reproductive rate of bobwhites, pheasants, and mice; apparently, such doses also reduce the resistance of certain birds to various kinds of stress. The United States Fish and Wildlife Service has reported finding traces of DDT in 31 out of 32 specimens of bald eagles they examined. This has led to a widespread belief that the near-extinction of our national emblem is related to reduced reproductive success resulting from the concentration of DDT residues in these large predators.

In the fall of 1963 there was much discussion in the papers about the death of millions of fish in the Lower Mississippi River system, deaths believed to have been caused by another, even more powerful insecticide, endrin, that apparently entered the river as runoff from agricultural lands or as waste from an insecticide factory in Memphis. A decline in the shrimp catch in the Gulf of Mexico may also have been caused by insecticides from the Mississippi; shrimp are known to succumb to minute quantities of various pesticides. Of course, it is always difficult to prove what caused death in a particular case, since our waterways are contaminated with such a variety of substances. Lake Erie is now largely a sterile lake—except for the algae that thrive on the wastes and wash in foul masses on the beaches—largely because it is ringed with cities that can find no better place to discard their wastes.

Insect-control technology is in fact becoming much more sophisticated than it was a few years ago. New chemicals are being tested with much greater care, and efforts are being made to synthesize substances toxic to only one or a few species and leaving no permanent residue on the plant or in the soil. Compounds similar to those produced by insects themselves seem especially likely to pay off in terms of effectiveness and nontoxicity to other organisms: substances such as synthetic hormones, pheromones, or repellent secretions. Carroll M. Williams, whom we met in Chapter 9, speaks of these as "third-generation pesticides." Lead arsenate and other "old-fashioned" insecticides comprised the first generation, DDT and its relatives the second. The reaction against the second generation is only partly attributable to the impact of *Silent Spring*. The fact is that nearly 200 species of insects are known to have developed strains resistant to DDT and other chemicals! As Professor Williams points out, an insect cannot develop resistance to its own hormones without committing suicide in the process. He believes that many substances found in plants and now of unknown function may in fact be found to mimic the juvenile hormone and to have been evolved by plants as means of deterring certain insects, as the "paper factor" he and Dr. Sláma found in the balsam fir apparently did.

Another good example involves the fire ant, as mentioned earlier the object of an expensive control program more effective in arousing sentiment against insecticides than in eliminating the ant. While this program was under way, E. O. Wilson, a colleague of Williams at Harvard, was quietly studying the behavior of the fire ant. He found that worker ants lay "odor trails" from sources of food by depositing a glandular secretion from the tip of their sting. Efforts are now under way to isolate and identify the chemical, and if these succeed it may be possible to synthesize a related substance and use it in control of the ant. ("Trail substance" itself disappears within a short while, and has to be renewed by the ants, so it may be necessary to develop a more persistent analogue.) Very likely the substance will prove nontoxic—even to ants—but its application may result in utter confusion and eventual starvation of colonies of the fire ant.

These are only two examples of new approaches—two that happen to have been developed close to me. Others are to be found in the "sterile male" technique mentioned in Chapter 8, the spreading of insect disease organisms, and the manipulation of the environment so as to prevent population build-ups (for example, by the desert locust). It is fortunate that a few good entomologists have been engaged in fundamental research in these very practical times, for basic, "curiosity-oriented" research seems again to have provided the background for new technologies—desperately needed in the field of insect control.

Biological control is, of course, a field of potentially great importance, though in fact it represents one of the older approaches. C. V. Riley, you will recall, used subterfuge to bring in the Australian lady beetle to control the fluted scale, and he was awarded the Legion of Honor for suggesting that French wine producers graft their plants onto American rootstocks, which were resistant to phylloxera. For a good many years we have been using varieties of wheat bred for resistance to rust and to the Hessian fly. And the importation, rearing, and release of parasites continues, at least in some areas. In this field, *Silent Spring* has perhaps had the unfortunate effect of raising hopes beyond present justification. Many persons seem to believe that simply by stopping the application of insecticides one can expect the parasites eventually to take over. On the Harvard campus, the ivy is seriously attacked every year by the caterpillars of the eight-spotted forester, an attractive little moth. For a number of years the ivy was sprayed regularly and the caterpillars reasonably well controlled. But after robins were seen to be made ill by eating poisoned caterpillars, the spraying was stopped on certain buildings, on the assumption that one or more of the known parasites of the insect would take over. The result has been almost total defoliation that may go on year after year until the ivy dies (as some already has). The fact is that it has been a very long time since any part of Cambridge, Massachusetts, was a "natural area" in any sense of the word, and it is unrealistic to expect the "balance of nature" to be in operation there. Conceivably the introduction of parasites might work, but it has not been tried. Perhaps there is a moral to be drawn: that ivy doesn't grow on towers that are too much of ivory.

Insect control by the use of parasites and predators is surely the most satisfying method to contemplate, and fortunately there are several centers of intensive research in this field. But the field is an exceedingly complex one, and on the whole there is little evidence that we are willing to pay for the large amount of fundamental research necessary on the systematics and biological relationships of parasitic wasps, flies, and other arthropods. The development of "third-generation pesticides" remains, for the moment, a hopeful prospect for the future. Right now we must face up to the fact that the method of control that is least satisfying intellectually—the use of chemical poisons—is the only one broadly applicable to the vast acreages of high-yield crops being grown to feed an increasingly crowded world.

The fact is that the pesticide situation is but one small facet of what man is doing to his planet in this era of unprecedented expansion. I was interested to discover that in the 1,193-page symposium published by the University of Chicago in 1956, and titled *Man's Role in Changing the Face of the Earth,* the word "insecticide" appears three times! The essence of this book is expressed in a nutshell by one of the contributors, Chauncey Harris: "We stand today in the midst of a gigantic and pervasive revolution, the urbanization of the world."

It is said that urban and suburban developments in the United States expand over about three million acres of land each year. In California, bulldozers are said to level an average of 375 acres a day. I have no figures on the matter, but I venture to guess that for every songbird killed by DDT, a thousand are permanently routed by the bulldozer.

The matter with respect to wildlife is summed up nicely by R. L. Rudd and R. E. Gennelly, of the California Department of Fish and Game:

"Considered in its broadest scope, at the present time, pesticides seem to be only minor influents in nature compared to other factors in land and water development and use. Urbanization, industrial pollution, drainage of marshlands, bringing land into cultivation—to name a few such factors—all constitute greater hazard to wildlife than chemical use."

This is what is often called "the revolution in the environment," a revolution that is partly a consequence of our expanding technology, partly a consequence of mere increasing human numbers—in itself, of course, a result of the influence of improved technology on death rates. Each hour the world's population increases by about 5,000 people; each week a city the size of Cleveland, Ohio, is added to our planet; the population *increase* between now and the year 2000 will exceed the present population of the earth. Although the rate of population increase has been declining slightly in the United States, at the present rate the population will still double within our childrens' lifetimes. In Latin America, where the population growth rate is the highest in the world, there will be nearly three times as many people by the year 2000! Alarm over the impending starvation and economic ruin implied by these statistics led to the first inter-American conference on population policies, held in Caracas in September, 1967. Such a conference would have been unheard of only a few years ago, though in fact it confined itself largely to planning for the increase—birth control still being a forbidden subject in many political circles.

Not only will there be many more people, but those people will hope to share more fully in the earth's bounty—thus further augmenting the revolution in the environment. In our affluent society, the per capita production of trash has risen from 2.75 pounds per day in 1920 to 4.5 pounds per day in 1965. One million people (in an industrialized society) produce 500,000 tons of sewage a day and discard 1,000 tons of hydrocarbons, carbon dioxide, and other substances into the air. Each day 16,000 automobiles are scrapped, to say nothing of cans, bottles, waste paper and plastic products. According to the Los Angeles *Times*, the residents of that city are submersed each day in nearly 9,000 tons of carbon monoxide and 1,180 tons of hydrocarbons, as well as sizable amounts of ammonia, lead, and sulfur compounds. In New York, the mayor's Task Force on Air Pollution recently reported that if that city had the sheltered topography of Los Angeles, everyone "would long since have perished from the poisons in the air."

A recent press release notes that Lake Michigan receives untreated sewage from 55 communities and partially treated sewage from 88 others. Chicago's wastes are normally sent south via the Illinois River, but during a heavy rain in the spring of 1967 the system was so overtaxed that 400,000 tons of raw sewage were discharged into the lake to prevent flooding in the city—forming a mass of sludge 2 miles wide and 15 miles long. In addition, Lake Michigan is ringed with industries that contribute great quantities of wastes. The Calumet River, just south of Chicago, carries more than 35,000 gallons of oil into the lake each day. The Gary plant of United States Steel alone contributes 16 tons of iron waste and 280 pounds of cyanide each day—and it is only one of several such plants in the area. Industries in the Milwaukee area are said to add (among other things) more than 9,000 pounds of phosphorus per day. Lake Michigan, like so many of our lakes and streams, has become a cesspool for our affluent society—or *effluent* society, as many persons are coming to call it. The committee on pollution of the National Academy of Sciences has pointed out that at present rates of increase there will be, by 1980, enough waterborne waste to consume all the oxygen in all twenty-two river systems in the United States. In the words of the committee:

"Pollutants are the residues of the things we make use of and throw away. . . . As the earth becomes more crowded there is no longer an 'away.' One person's trash basket is another's living space."

As former mayor of Cleveland Ralph Locher puts it: "We could be known as the generation which put a man on the moon while standing ankle-deep in garbage."

A recent survey reveals that most persons believe that man's greatest challenge lies in finding a cure for cancer, with the attainment of a permanent end to war ranking second. It is ironic that we place so much emphasis on survival and so little on the kind of world we wish to survive in. We know that we have rendered many of our waterways sterile and foul. We know that smog can have serious effects upon health. We also know that our burning of fuels over the last century has caused a 14 per cent increase in the amount of carbon dioxide in the atmosphere. This has already caused a slight general increase in temperature, and it is postulated that by the year 2000 global temperatures may have increased by as much as 4 degrees as a result of the "greenhouse effect," that is, the overlay of carbon dioxide preventing the escape of heat into space. This may have strange effects on weather patterns, and furthermore may melt parts of the antarctic ice sheet and thereby cause a rise in ocean levels—even to the extent of permanently flooding most of our coastal cities.

"Noise pollution" is perhaps of a less serious nature, but there is no doubt that we live in an exceedingly noisy world. Can we make the decisions called for; can we keep our heads in the atmosphere of roaring, grinding, and shrieking with which we surround ourselves? Springs in my neighborhood are silent of birdsong, yes, but far from silent! The quiet buzzing of the hand lawnmower has been replaced by the piercing whine of the power mower. Trucks roar down the superhighway not far away; helicopters clatter over our rooftop, and jets bumble their way to and from the local airport. But I do not complain, for I know that all this will seem like the sweet joys of yesteryear when the giant supersonic jets are here! Just recently I visited several of our National Parks in the Rocky Mountain area. Not one of them, in all their pristine glory, was free from that inescapable reminder of human "progress": sonic boom. As I looked up at gigantic rocks perched on top of the great cliffs of Zion Canyon and remembered reports of buildings collapsing as a result of sonic boom (as well as pinnacles toppling in nearby Bryce Canyon), I wondered again about this matter of balancing benefits against risks.

Barry Commoner, in his recent book *Science and Survival,* has well expressed the fears of many thinking men:

"We have come to a turning point in the human habitation of the earth. The environment is a complex, subtly balanced system, and it is this integrated whole which receives the impact of all the separate insults inflicted by pollutants. Never before in the history of this planet has its thin life-supporting surface been subjected to such diverse, novel, and potent agents. I believe that the cumulative effects of these pollutants, their interactions and amplification, can be fatal to the complex fabric of the biosphere. And because man is, after all, a dependent part of this system, I believe that continued pollution of the earth, if unchecked, will eventually destroy the fitness of this planet as a place for human life."

The study of the relationships of organisms to their environment is called ecology. Ecology is a new science; the word itself is less than a century old, and only in the last few decades has the subject become a respectable one for a scientist to pursue. Human ecology is even newer, but suddenly it has emerged as a popular subject for books and symposiums, indeed, as a major theme in current thought. Britain has established a Natural Environment Research Council, which has an initial budget of 3.8 million pounds for ten months. The American Association for the Advancement of Science devoted its 1966 meetings in Washington, D.C., to the theme "How man has changed his planet"; and the American Institute of Biological Sciences held a plenary session at its 1967 meetings in College Station, Texas, on the subject "Environment of Man Revisited." Courses and research programs in human ecology have been established in several leading universities: for example, at the University of Missouri School of Medicine, the Harvard School of Public Health, the Cornell Medical Center, Boston University, and the University of Illinois. It does indeed seem that man has entered what Secretary of the Interior Stewart Udall has called "the era of ecology." In Udall's words:

"Francis Bacon once observed 'We cannot command nature except by obeying her.' The wise import of these words has been obscured during the budding spurt of our technology. In the flush of first knowledge, we have imagined that we could control without obeying. As we survey now the results of waning wetlands, persistent pesticides, smog-filled air, and troubled water, we are facing the realization that only by working with nature in her ecological entirety will man realize his highest potential in the scheme of things. Ethical, esthetic, and utilitarian reasons all support the efforts to conserve the diversity of nature. This is good and right for the diversity will guarantee future generations the richness and satisfaction which a narrower, more man-centered universe would deny them."

There are, of course, many aspects to this problem, but as a biological taxonomist and curator I am particularly concerned with this matter of diversity, which Udall stresses. The fauna and flora of the world are rapidly becoming simplified and standardized. This is not primarily the fault of wanton hunters and hard-fisted lumber barons; it is a consequence of man's motility and his efforts to build a "better" life for himself wherever he goes. As Charles Elton points out in his book *The Ecology of Invasions by Animals and Plants,* each continent was once occupied by its own distinctive plants and animals. But man, both intentionally and inadvertently, has carried great numbers of them to other continents. So far as this results in an enrichment of the natural world, no one can complain, but time and again the invader, free of its natural enemies and often more "vigorous" than the natives, undergoes a population explosion and becomes a "pest"—like the prickly pear cactus in Australia, like the starling, the European cornborer, the Dutch elm disease, and a great many other organisms here in North America. Those plants and animals that are able to thrive in the presence of man are coming to dominate the landscape. As I look out my own window I see house sparrows, pigeons, cats and dogs; among the insects, honeybees, cabbage butterflies, and Japanese beetles—a "nature" largely devoid of anything native. This is a mark of our times, and we are fortunate indeed that some organisms do adapt to our presence. Many do not, or at least are not able to compete with our many introductions.

According to Elton, the United States Office of Plant Introduction has been instrumental in introducing nearly 200,000 kinds of plants to this country from all over the world. In the course of such international shipments of plants, many of our major insect pests were introduced. In the effort to control some of these, we have deliberately introduced more than 400 parasitic and predatory insects to the continental United States (although not all became established). If this influx of new organisms of all kinds continues unabated, all life on earth will eventually be homogenized and strained through the filter of adaptability to man and his managed crops and forests. Of course, there will be latitudinal differences: the mongoose, an introduced carnivore that has decimated the snakes and small mammals of the West Indies, is not likely to become part of the fauna of the northern United States, nor will the gypsy moth become a pest of coconut plantations. This homogenized, human-adapted fauna will itself undergo changes as we change our food sources and living habits. The body louse and the bedbug, for example, have been largely eliminated in the more affluent sectors of man's world; their ultimate fate will depend upon the eventual distribution of the world's limited resources among the peoples of the world. If all live under highly crowded conditions and at a mere subsistence level, and if they fail to find solutions to problems of pollution and waste disposal, there will be a great future for the bedbug, the louse, the filth flies, and the rat, to say nothing of various bacteria and viruses. On the other hand, if man is capable of planning his future, and on a global scale, it is probably within his power to include, with this homogenized fauna, squirrels instead of rats, butterflies instead of bedbugs.

That we are swiftly achieving a limited, man-adapted fauna and flora over much of the world cannot be denied. The vast herds of hoofed mammals that once roamed Africa have been largely replaced by cattle; at least two species are said to have recently become extinct, and several others occur in greatly reduced numbers. The wonderful marsupial fauna of Australia has been ravaged to make way for sheep, cattle, and rabbits. In the recent book *The Great Extermination,* edited by A. J. Marshall, six species of marsupials are listed as extinct, twenty-eight others as endangered—though some on the "endangered" list, such as the Tasmanian wolf, may in fact be extinct. Marshall's account of the slaughter of the kangaroos, wombats, and koala bears makes frightening reading—in one month of 1927, 600,000 koala bear skins were marketed in Queensland, and this prototype of the "Teddy bear" now survives only in a few parks, where it is protected as a tourist attraction. The growing importance of the tourist industry, incidentally, provides the last hope of survival for a great many of the world's larger and more attractive or bizarre animals.

Of course, North Americans are in no position to sneer at the Australians or the Africans. We simply began homogenizing our fauna sooner and have tended to brush the matter under the rug along with our treatment of the original human inhabitants of the continent. Peter Matthiessen, in his book *Wildlife in America,* has recently been looking under that rug with respect to the birds and mammals, some fifty species of which are presently regarded as "biologically endangered." As Matthiessen puts it:

"The slaughter, for want of fodder, has subsided [but] such protection as is extended them too rarely includes the natural habitats they require, and their remnants skulk in a lean and shrinking wilderness. . . . The true wilderness . . . the great woods and clear rivers, the wild swamps and grassy plains which once were the wonder of the world—has been largely despoiled, and today's voyager, approaching our shores through the oiled waters of the coast, is greeted by smoke and the glint of industry on our fouled seaboard, and an inland prospect of second growth, scarred landscapes, and sterile, often stinking rivers of pollution. . . . Where great, wild creatures ranged, the vermin prosper."

The vertebrate animals, especially the birds, at least have eloquent defenders, although often their voices are carried away on the winds of "progress." The insects and other smaller creatures of the earth seem to have few advocates—despite the overwhelming importance of the lower strata of the "pyramid of numbers" in supporting the upper. I am glad to report that a couple of years ago the Entomological Society of America established a committee on "Insects in Danger of Extinction." The committee found that most cases of near-extinction were associated with destruction or alteration of the habitat.

The most striking examples of the replacement and elimination of native plants and animals are to be found on islands remote from continents. On most such islands various unique organisms have evolved, and they have evolved to fit the particular conditions prevailing in some habitat on that island. When plants and animals are introduced from continental areas, they almost invariably replace the natives. This can be seen dramatically in New Zealand, for the settlers of that island attempted to re-create their native England, and in so doing brought in nearly 50 kinds of mammals, over 100 kinds of birds, and more than 500 kinds of plants. Not all became established, but some flourished beyond all expectation. Imported watercress ran wild, and developed stems as thick as a man's wrist, choking many of the rivers. Hordes of wild pigs and goats ran rampant over the countryside, clearing the vegetation as only they can do, and rendering many of the native birds and mammals literally "scarce as a kiwi." As a specialist on wasps, I have been especially interested in the remarkable success of the European yellow jacket in New Zealand. It first became established about 1944, perhaps from hibernating queen wasps emerging from crates containing airplane parts. Within four years it was so tremendously abundant that a bounty was paid for every queen sent in to the Department of Agriculture. This resulted in the collection of 118,000 queens in three months, but there was no noticeable reduction in the population the next year. Although colonies do not survive the winter in Europe (the queens hibernate and found new colonies in the spring), in New Zealand they do often survive the winter and grow into even larger colonies the next year. One yellow-jacket nest fifteen feet high and five feet wide was discovered—a fantastic size for a wasps' nest. Like most other wasps, this species is a predator, but its effect on the native insects will remain largely unrecorded, since in fact the insect fauna was vastly altered by man before it was ever studied in a thorough manner.

Even more catastrophic changes have occurred on smaller islands, especially in archipelagoes in which each island has developed its own particular assemblage of plants and animals. It was the study of such archipelagoes—especially the Galápagos, in the case of Charles Darwin, and the East Indies, in the case of A. R. Wallace—that provided some of the first insights into the processes of evolution. One of the most interesting archipelagoes is now one of the United States, and therefore especially worthy of our attention. The Hawaiian Islands are two thousand miles from the nearest continent, and have evolved a rich and unique fauna that originally shared only a very small percentage of species with other areas. Furthermore, the fauna of each of the major islands is in some measure unique. One entire family of birds, with twenty-two species, occurs nowhere else in the world: the drepanids, or sicklebills. In fact, only a few of these birds have sickle-shaped bills (for extracting the nectar of flowers); others have bills adapted for feeding on fruits or on insects. Apparently the original ancestor of the group found few other birds on the islands, and evolved into diverse types that took advantage of available sources of food. Yet these remarkable birds, which have provided such a wonderful example of adaptive radiation, are now rare and confined to forested areas in the mountains. They have not been able to compete with importations such as the mynah bird, and may have succumbed to bird malaria and other introduced diseases. Ground nesters have suffered from depredation by mongooses, rats, and pigs. Cats, too, have long run wild on all the islands, and needless to say have not had a beneficial effect on the bird fauna. Well over half the native bird species are now extinct.

In the introductory volume of his monumental *Insects of Hawaii*, Dr. Elwood C. Zimmerman wrote (in 1948) that he believed that a third or more of the native species of insects may have become extinct since the arrival of man. Many of them were restricted to feeding upon some particular plant, or were parasites of one particular insect, and they disappeared when their hosts were eliminated. Others have fallen prey to the notorious big-headed ant, Pheidole megacephala, a species that has caused great alterations in the insect fauna on many islands to which it has been carried by man. According to Zimmerman:

"One can find few endemic insects within the range of that scourge of native insect life. It is almost ubiquitous from the seashore to the beginnings of damp forest. Below 2,000 feet few native insects can be found today, and those which are found there belong to a few species which form an unusual small assemblage of forms which have been able to withstand the changing environment, or have adapted themselves to new hosts."

Yet the Hawaiian insect fauna was one of the most fascinating in the world. Zimmerman calculated that the approximately 3,700 species of insects known to be native to Hawaii evolved from no more than about 250 immigrants (most probably via air currents and island "stepping-stones" from the western Pacific). Several immigrants "went wild" on the islands, proliferating into large groups. One genus of moths and one of flies each contains over three hundred species on this small archipelago, while one genus of beetles and two of wasps each contain over a hundred. One of the latter is a group of solitary wasps of considerable interest to me. All prey upon caterpillars, but the fact is that so many parasitic wasps have been deliberately introduced into Hawaii for the control of pest species that most of the native caterpillars have been greatly decimated. It may be, too, that parasites and diseases of the wasps themselves have been inadvertently introduced. At any rate, most of these solitary wasps are now exceedingly scarce, and some will surely be gone before the group receives the critical study it deserves.

Still, this is apt to be of little concern to most persons. The average citizen of Hawaii, like anyone else, is interested in progress. The average tourist is apt to prefer mynah birds to sicklebills, and is not unhappy to learn that the native wasps are dying out. Yet a person interested in more than superficialities is likely to regret the fact that Hawaii, Puerto Rico, and the Canary Islands are coming to share pretty much the same plants and animals, just as one part of the temperate region is coming to look pretty much like another. If the future finds us with more leisure and more opportunity to travel (as we all hope it will), we may find it harder to escape boredom than we anticipate.

A group of islands now much in the news is Aldabra, a coral atoll 260 miles northwest of Madagascar. Of its twenty-two species of land birds, over half are found nowhere else, and many of the plants and insects are unique. It is the last home of the giant land tortoise in the Indian Ocean, and green and hawksbill turtles breed there in numbers, to say nothing of countless numbers of sea birds. The British and United States Air forces now want to convert the atoll to a base for military jets—and the islands are much too small to support a major airfield without almost total destruction of the native plants and animals. Biologists are, of course, rushing to the defense of Aldabra. As of this writing it appears that the atoll has been saved (at least for the moment), not by the biologists but by reduced public spending following the devaluation of the pound.

A person who has never visited a small island thrown wholly out of balance by the hand of man can scarcely appreciate the enormity of the changes. When Elwood Zimmerman arrived in 1934 on Mangareva, an island in southern Polynesia about three hundred miles from Pitcairn, he and his co-workers expected to explore a rich and novel fauna.

"But [wrote Zimmerman] what a bitterly disappointing place we found Mangareva to be! All the native forests are gone—burned and reburned, and eaten away by goats. After scouring the islands, we found a few native plants on a steep cliff, and on these we found a few endemic insects. . . . Today on Mangareva, nothing but the ghosts of a once unique and magnificent biota hover—their cries are echoed by the moaning birds screaming over the mountain slopes barren of native life and mocked by the bleating of hungry goats. . . .

"We can hardly guess at what the lowland biotas of vast island areas in the inner Pacific were like before the coming of man. . . . Each island has its characteristic biota which is the end product of hundreds of thousands or millions of years of evolution. Each of these biotas has developed only once in the history of the world. Once lost, they can never be regained. They are extremely delicately balanced, and chain reactions of extermination processes may be set off by what may seem to be minor disturbances. The destruction of a single element may have long-reaching effects."

It is too late to save Mangareva, too late to save much of Hawaii. Human curiosity and compassion should themselves provide a sufficient basis for preserving what is left of our plundered planet. But there are still more cogent reasons for trying to preserve the diversity of nature: How do we know a given species may not be the key to our understanding of some problem, or may not provide some unknown benefit? Who could have supposed that a puny little fly, Drosophila, would teach us so much about genetics? Who could have supposed that we would try to recall the manatee from near-extinction to clear our inland waterways—since one manatee will eat up to 100 pounds of water hyacinths per day? Who could have supposed that a green mold contaminating a culture of bacteria would lead us to the development of antibiotics?

Just recently a chemist in Heidelberg named Schildknecht discovered that the glands of an obscure water beetle contain cortexone in amounts equivalent to that in the adrenal glands of more than 1,300 cattle. Cortexone is important in the synthesis of certain medicines, and currently sells for $6.50 per 100 milligrams. In pointing this out in a recent essay, Robert L. Metcalf, of the University of California in Riverside, comments that the insects have "become the 'new frontier,' for natural products chemistry." After all, they require very little in the way of space and food in the laboratory, and considering the size and diversity of the group, who can predict what we may still gain from them? Anyone familiar with the history of biology knows that every breakthrough has been the result of a person with the right background working on the right organism at the right time. As Gene Marine puts it in his article "America the Raped":

"We have no way of knowing what characteristics of what animal, plant or microbe may someday prove to be in some way valuable. . . . A byproduct of the whooping crane may be tomorrow's wonder drug. The ecology of [an] estuary may provide the clue that enables us to project a more viable ecosystem for a space station."

I do not see how this argument for maintaining the diversity of nature can be reasoned away. One may deny, if he wishes, the aesthetic and spiritual values of nature, but he cannot deny that man owes all that he is to the world that has nurtured him—or that we still have a long way to go in probing its depths and reaping its benefits. It is unfortunate that our population and technological explosion have come at a time when ecology—the most complex and difficult of all the sciences—has not yet had time to build a truly convincing and coherent body of facts and theory. We hear of the need for an "ecological conscience" or, as Aldo Leopold called it, "an ethic of the land," yet the ecologists often disagree among themselves or are too shrouded in gloom to have anything constructive to say. It is perfectly true that ecology is the most underdeveloped and undersupported of all the sciences in relation to what we still have to learn. What is more complex than the total environment of an organism: all the plants and animals that impinge upon it at one time or another, whether as food, competitors, parasites, predators, or whatnot; all the nonliving elements in the environment, such as light, soil type, wind, rainfall, temperature and so forth? Each member of a natural community is interdependent upon all the others, and each is influenced in its own way by climate and other physical factors. And communities change with time: in the course of the season; over the years, as when a forest encroaches upon a clearing; and over millenniums, as the members of the community evolve their adaptive adjustments to one another and to the climate.

Ecology rests upon descriptive natural history, that is, upon a knowledge of all members of a community (or "ecosystem," in the current jargon): what these members are, what they feed upon, where they make their homes, when they are born and when they die, and so forth. As I have said many times, it will be a long while before we become fully acquainted with all the insects and other small organisms that inhabit the earth. Nevertheless, ecologists, building upon partial knowledge of a few natural communities and laboratory experiments with highly simplified situations, are beginning to provide us with important insights into the nature of nature.

Most natural communities, consisting as they do of a host of plants and animals that have evolved together over a long period of time, tend to be relatively stable. If some climatic variation favors one species for a time, it is soon checked by others. This is what is sometimes called the "balance of nature," although this term conjures up a simpler and more static system than usually prevails. Introduction of a vigorous species from a similar but distantly removed community may have major consequences at least until the members of the community are able to adapt to the intruder. A gross change in physical factors—for example, a new layer of silt deposited by a flood or the burning over of a forest and consequent increase of light—may cause widespread local extinction and the development of wholly new communities. It is man's tendency to change the environment radically, perhaps even more than his propensity to carry organisms into new regions, that produces such profound effects upon the complexion of nature. A bulldozer is a factor that brooks no compromise or adjustment.

Study of natural communities reveals that those occurring in rigorous climates or substrates—for example, in the arctic or in hot springs—tend to have only a few species. These are the few that have been able to adapt to these difficult habitats, where they have few natural enemies and often occur in enormous numbers—for example, the mosquitoes of the far North or the weeds that fill a field stripped of its topsoil. Mark Twain, in his book *Roughing It,* described in memorable words the plight of a California lake he characterized as being "nearly pure lye":

"There are no fish in Mono lake—no frogs, no snakes, no polliwogs—nothing, in fact, that goes to make life desirable . . . [nothing] except a white feathery sort of worm, one half an inch long, which looks like a bit of white thread frayed out at the sides. If you dip a gallon of water, you will get about fifteen thousand of these. . . . Then there is a fly, which looks something like our house fly. These settle on the beach . . . and any time, you can see there a belt of flies an inch deep and six feet wide, and this belt extends clear around the lake—a belt of flies one hundred miles long. . . . You can hold them under water as long as you please—they do not mind it—they are only proud of it. When you let them go, they pop up to the surface as dry as a patent office report, and walk off as unconcernedly as if they had been educated especially with a view to affording instructive entertainment to man in that particular way."

Mark Twain was not exaggerating. Photographs of great windrows of fly puparia have been published (for example, by E. O. Essig, in his *History of Entomology*). Twain was apparently not aware, however, that the flies were the adult stage of the "worms" in the lake. LaMont C. Cole quotes Mark Twain's comments on Mono Lake in his provocative essay *The Impending Emergence of Ecological Thought,* and then adds:

"The principle illustrated is of wide generality. Arctic waters contain fewer species than tropical waters, but far more individuals of each. The same applies to tundra as contrasted to the tropical rain forest. This trend can be traced from cold springs to thermal springs and from low-altitude to high-altitude situations. And genuinely difficult environments, not only saline lakes but also sewage beds, sulfur waters, and the like, produce almost incredible concentrations of one or very few species. . . ."

Here, then, is an ecological principle that strikes close to home—one that we see functioning when a heavy dose of DDT on an orchard creates a "difficult environment" for most organisms—but allows orchard mites to burst forth in a great population explosion; or when a lake become so heavily polluted that it dies—except for the algae that

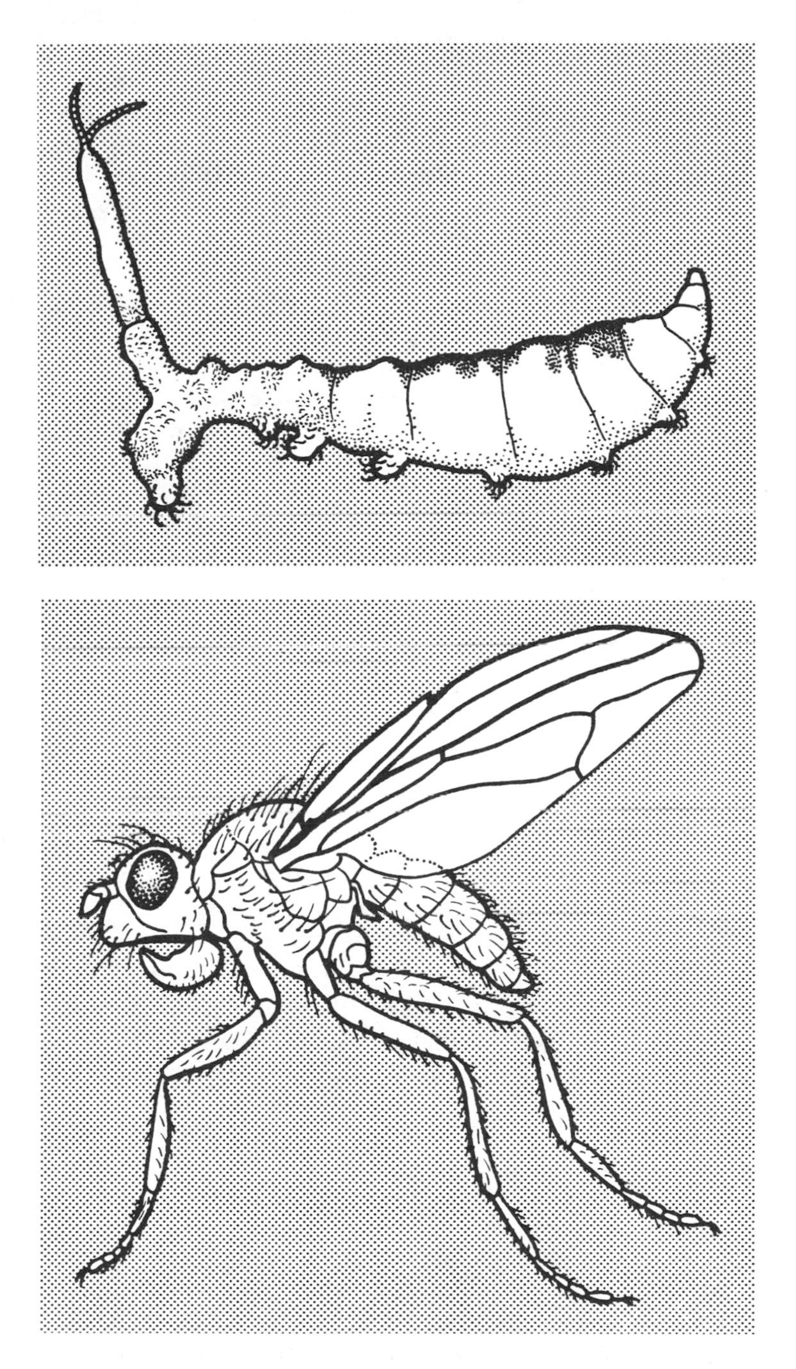

The larva and adult of the fly of Mono Lake, California. The local Indians, who called it the koo-tsabe, used to harvest the pupae in great numbers in the fall and use them as food.

breed in fantastic numbers. Admittedly the retention of diversity is sometimes difficult in our modern world, but if we are mindful of the values of diversity, of the sacredness of life—and I do not see how one can sanely restrict this to human life—then there are surely things we should do and things we should not do. In our personal affairs we are guided by our consciences, by our awareness that we are dependent upon the rest of the human community and must act accordingly. Man as a species must develop an awareness that he is infinitely enriched by the organic world, that he is himself poorer each time a species is eliminated or a natural community permanently disrupted.

It would be nice to be able to report that basic problems in ecology are occupying the time of the majority of biologists. In fact this is very far from true. Most ecological problems require intensive research on many levels over a period of years. They require accurate identification of species (and there are many hundreds in most natural communities); they require knowledge of what these species do and how they react to one another and to temperature, light, day length, and so forth; they require sampling of populations over many years in order to detect oscillations and interactions; they require experimental manipulation and sophisticated mathematical treatment to bring out trends and key influences. The use of computers may soon make much more penetrating analyses of ecosystems possible. A great many persons may be required to feed the computers with data gathered from nature or from experiments, but this would be a more worthwhile use of human energy than some I can think of. It might even turn out to be a way that persons of moderate education can join the fun of science.

Unfortunately, we live in a time when funds to support long-term projects with no guarantee of "useful" results are virtually nonexistent. Colleges and universities rely greatly upon funds from government sources, and these are invariably doled out on a one-, two-, or at most a three-year basis. State-supported universities are a good deal better off than private ones, but state funds are almost always shunted into projects likely to benefit the citizens of that state this year or next. I speak from experience, having once resigned a position in a leading agricultural college because the head of the department (now a dean!) insisted on using "benefit to the farmers" as a common denominator for every transaction. Also, I have seen state after state conduct a similar project on, let us say, the insecticidal control of the hairy-snouted quince-girdler, when not one of them has a long-range project on the ecology of orchards and the side effects of pesticide applications upon them. "Cost benefit analysis" and "mission-oriented research" are the order of the day, and even the National Science Foundation is evidently to be reorganized so as to permit it to support applied research.

Even within the scientific elite there is no unanimity on such matters. The *Reader's Digest* recently quoted Dr. Paul Weiss of Rockefeller University to the effect that "the fetish of unfettered research" should be replaced with the maxim that scientific inquiry "should have a purpose and be selected with a sense of relevance." A recent editorial in *Bioscience* admonishes biologists for lacking social responsibility, and calls for "more effective mobilization of research toward the solution of pressing problems." The editorial also calls for biology courses that give students "a better understanding of the role of biology in solving some of our pressing problems." This is fine, but if we teach them that "mobilization" is the whole answer, we shall be committing a serious error. Most of what we know of nature was gathered by persons—many of whom would have to be ranked as "amateurs"—who were *curious*. If we relegate curiosity to a subordinate role, we shall sterilize science.

Comments about "relevance" and "the solution of pressing problems" strike me as only a slight improvement over the well-known comment of a former Secretary of Defense that basic research is "when you don't know what you're doing." Basic research is simply modern lingo for satisfying one's curiosity. It is indeed "mission-oriented," and its mission is to add to man's understanding of the world, and incidentally to provide the facts and concepts that will enable him to build a richer life. The history of science is the story of a constant flow of useful information from the vast coffers of basic research into human affairs. In these crisis-ridden times it would seem wise to strive to fill these coffers furiously, for we are in urgent need of whatever new ideas and insights may be drawn from them.

It is only by accumulating facts about a great diversity of things—most of them insignificant in themselves—that we can begin to see relationships and to find patterns that will guide us. Who can say what piece will be the clue to the puzzle, what feature of what plant or animal will be just the information we needed all along? It will be surprising to me, incidentally, if all this "zeroing in" we hear about in the field of medicine will produce a cure for cancer. That will come of itself when, in some curiosity-oriented laboratory, someone has probed deeply enough into the basic phenomena of growth—who knows on what obscure creature?

The realization that nothing less than knowledge in depth of the totality of life on earth must be the ultimate goal of biology—the ultimate science—does not frighten a curious person; it merely stimulates him. What frightens him is the difficulty of explaining to a non-biologist why this must be so. My own immodest solution (which I shall save for another chapter) is to convert the world to biology.

Is Nature Necessary?

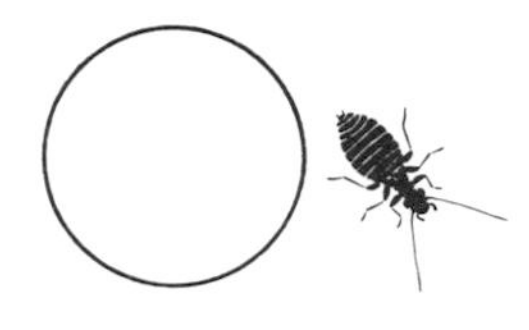

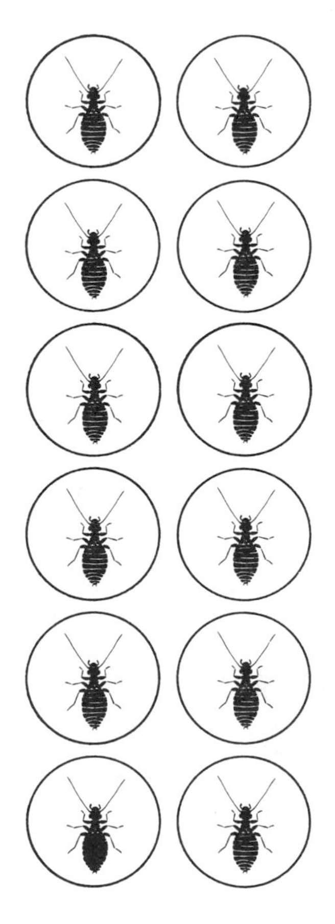

A few years ago a gentleman came up to me when I was mounting wasps at a picnic table in a Missouri state park. "What is the purpose of a wasp?" he asked. Had I been a lepidopterist, he doubtless would have asked the purpose of a butterfly, though I'm not sure what he would have asked had I been an anthropologist. I don't recall how I answered his question, but I'm fairly sure he didn't find my answer satisfactory. I was reminded of Dylan Thomas' listing of "useful presents" in *A Child's Christmas in Wales:* "mittens made for giant sloths; zebra scarfs . . . and books that told me everything about the wasp, except why."

In *The Firmament of Time,* Loren Eiseley mentions a young man who asked him: "Why can't we just eventually kill off everything and live here by ourselves with more room? We'll be able to synthesize food pretty soon." Admittedly there are few who would want to go to this extreme or who would insist that everything in nature serve some easily defined utility in man's world. Yet scarcely anyone would go to the other extreme, either; no one wants termites in his house or grizzly bears in his back yard. Man's history has largely been one of learning to live more fully and in greater numbers by taking advantage of those forms of life that had something to offer him and by disposing of those that did not. Now that he has gone so far and altered the natural world so vastly, perhaps it is relevant to ask whether life in a wholly or even mostly artificial world is possible or desirable.

In fact, is nature necessary to man? Will it not be a finer world when the last mosquito has been eradicated, the snakes relegated to books on extinct reptiles, the native forests replaced with fine, nursery-grown trees? Is this not already approaching fulfillment in Megalopolis, and is it not likely that it will be universally true in pages of man's history soon to be written? Should this be a matter of concern to us? Obviously I think it should. But I was brought up on a farm, with forty acres of cropland and another twenty of woods and meadows. That farm is now a housing development, complete with supermarket, laundromat, and even a superhighway passing approximately over our asparagus patch. Will the hundreds of children growing up on that sixty acres feel the way I do about these things? As man becomes more and more densely packed, more and more urbanized, we can be sure his attraction will be to the crowded, gadget-filled world that surrounds him rather than to the fireflies, killdeers, and trailing arbutus that used to entertain me as a child (and still do).

Even Aldo Leopold, that most articulate of conservationists, expressed doubts that there would be a place for nature in tomorrow's world. In one section of his classic *Sand County Almanac* he describes the joys of canoeing on the Flambeau, a river in northern Wisconsin. A dam was to be built on the Flambeau, and Leopold says:

"It seems likely that the Flambeau, as well as every other stretch of wild river in the state, will be harnessed for power. Perhaps our grandsons, having never seen a wild river, will never miss the chance to set a canoe in singing waters."

Leopold characterized the effort to preserve natural areas as no more than a "rear guard action" against the advance of civilization:

"Your true modern [he says] is separated from the land by many middlemen, and by innumerable physical gadgets. He has no vital relation to it; to him it is the space between cities. . . . Turn him loose for a day on the land, and if the spot does not happen to be a golf links or a 'scenic' area, he is bored stiff."

Aldo Leopold is surely right that man is growing apart from the world that nurtured him. Our loss of contact with nature is enhanced by lack of instruction in our homes and schools. Teaching of ecology or conservation in our lower grades, when it exists, tends to be oversimplified—the water cycle, the names of the trees and mammals, and so forth. Yet children have an intuitive liking for living things, and a desire to know what they do. Their attraction to insects, for example, could be greatly enhanced if they were taught the stories of their lives and of their interrelationships with other living things. High school and college biology courses often begin with biochemistry and rarely reach with any enthusiasm the whole organism and its relationship with its environment. There is nothing wrong with biochemistry, presently one of the most glamorous and well-supported fields of endeavor, but it is once or twice removed from the world of our experience. The term "nature study" has come to have a mushy, old-fashioned flavor, but its subject matter remains the best possible introduction to the study of the environment as well as to more sophisticated realms of biology—and I know some very good biochemists who started out watching birds or collecting butterflies.

If we are, in fact, witnessing the industrialization and urbanization of the world, and even talking about farming the seas, can we not be sure that such nature as survives will be limited and highly "unnatural"? Here we need to stop and think of what we mean by "nature." If we mean that which is wholly out of contact with humans, then of course nature is not necessary or pertinent to man. There may be some beautiful jungles on a planet encircling Arcturus, but if so they are of no present significance to us. Or we might go to the other extreme and claim that since man is part of nature so also is his culture. By this definition, the Beatles are as much nature as beetles, and one is as necessary (or unnecessary) as the other.

Of course, we don't usually define nature in either of these senses, but rather to designate those elements in the world that we do not directly manipulate. But here there is a difficulty, too, for man is above all a manipulator, and his mere presence soon results in alterations in natural communities. The hiker, the explorer, the student of forest, grassland, or aquatic ecology—even these have their effect upon wild nature. Many believe that generous samplings of natural communities need to be preserved relatively inviolate and open only to a few: this is the philosophy behind the Wilderness Act of 1964. There are real problems in knowing how small and circumscribed a wilderness area can be and how much human traffic it can bear without losing its integrity. Also, as the pressure of a growing population increases, it is going to be extremely difficult to keep these areas intact—many contain minerals, for example, that are in increasingly short supply. But for a moment the conservationists seem to have won a major legislative victory—provided the Act can be fully implemented.

There have been so many eloquent pleas for wilderness that I need devote little space to the matter here. While John Muir is rightfully considered the father of the wilderness movement, it is these ringing words of Henry Thoreau that are most often quoted:

"We need the tonic of wildness—to wade sometimes in marshes where the bittern and the meadow-hen lurk, and hear the booming of the snipe, to smell the whispering sedge where only some wilder and more solitary fowl builds her nest, and the mink crawls with its belly close to the ground. . . . We can never have enough of nature. We must be refreshed by the sight of inexhaustible vigor, vast and Titanic features—the sea-coast with its wrecks, the wilderness with its living and decaying trees, the thunder-cloud. . . . We need to witness our own limits transgressed, and some life pasturing freely where we never wander."

Thoreau's *Civil Disobedience* has become the manifesto of a generation disillusioned with the world it has inherited. There is hope that the rebellion against everything that standardizes and depersonalizes will include also a craving for Thoreau's "the tonic of wildness," for what Justice William O. Douglas speaks of as "the endless wonder and excitement of nature's flair for individuality rather than conformity." A remote canyon in the Rockies—or for that matter a close and appreciative look at a clump of grass in a city park—can serve some of the same ends as LSD, and with none of its side effects. The ease of travel now puts a variety of relatively unaltered natural areas at the ready disposal of everyone. Unfortunately the word *disposal* has a double meaning here, for the more we use these areas, the more their natural properties disappear. In the final analysis our concern must be with the complex interface between that remarkably successful creature called man and all other earthly organisms. Such an interface occurs everywhere—from city streets to the most remote wildernesses, even to the bottom of the deepest ocean trenches. The only reasonable view of conservation is that of Charles Elton, who has remarked that true conservation "means looking for some wise principle of co-existence between man and nature, even if it be a modified kind of man and a modified kind of nature."

So utterly different is the man-altered environment that Frederick Sargent and Demitri Shimkin of the University of Illinois have proposed calling it the "anthroposphere," in contrast to the original biosphere. They describe the anthroposphere as a series of artificial communities "designed to assure man of plentiful food and raw materials to meet his domestic and industrial requirements. The domesticated plants and animals . . . depend upon the continuing intervention of man for their survival. Because of competition with other organisms which continually invade these ecosystems man has been forced to institute control measures. . . . Furthermore . . . his artificial ecosystems are most susceptible to erosion through the agencies of water, wind, and fire. . . .

There has resulted what in essence is a continuing managerial struggle to maintain the artificiality of the domesticated landscape."

While the nature of the biosphere is largely determined by evolution, that is, by the types of organisms that have become adapted to specific temperature ranges, rainfall, and soil types—and of course adapted to each other—the anthroposphere is patterned to fit man's needs. It is here that the question—Is nature necessary?—really strikes home. How we build our cities and how we fill the space between them are problems of immediate concern to all of us. Do we wish our environment to be entirely artificial and managed, or shall we allow room for squirrels, ospreys, and crickets? What shall we save and what shall we destroy? Most of us would agree that a shaggy woodland, a sprinkling of wildflowers is good to have around, so long as it does not tie up land useful for economic reasons. But I happen to like yellow jackets; do I not have a right to yellow jackets if my neighbor has a right to a cat? In Aldous Huxley's *Brave New World* all have achieved perfect comfort and security, all but the Savage, who cherishes, among other things, "the right to be lousy." Although I have no affection for the louse, I am perverse enough to cherish the skunk in my woodpile, the silverfish in my bathtub. If freedom means anything at all, it means the right to choose one's environment and one's friends. I am an admirer of skunks, silverfish, and yellow jackets, and confess to a good deal of regard for the rattlesnake. If you were to press the point, I could probably think of something good to say about the louse.

I believe the strongest argument for keeping as much of the natural world as possible in the anthroposphere lies in the human need for variety, individuality, and the challenge of endeavoring to understand the nonhuman world. I believe, too, that emersion in a world of trees, flowers, and wild creatures is needed to nourish human attributes now in short supply: awe, compassion, reflectiveness, the brotherhood we often talk about but rarely practice except on the most superficial of levels. Last summer I met a student from the University of Maryland hiking alone high in the Teton Mountains of Wyoming. We were friends in a moment, eager to share our trials and enthusiasms; if either of us had needed help, it would have been automatic and unquestioning. Yet if we had encountered each other in our cars at a busy intersection in Washington, D.C., we surely would have acted with the usual rudeness and impersonality of city drivers. Human qualities grow best in an environment that transcends the human, that allows room for a stretching of minds and emotions.

Each year more and more Americans take their place in the endless rows of standardized homes that surround our urban centers, each home with its standardized nursery stock, lawn without dandelions, and so forth. So long as a golf course and supermarket are not too far away, the residents are reasonably content. Can the "Brave New World" be far?

While there is now much talk about an environment of beauty, purity, and diversity, there are cogent reasons why nothing much is done about it. As the suburbs encroach upon a given piece of land, the value of that

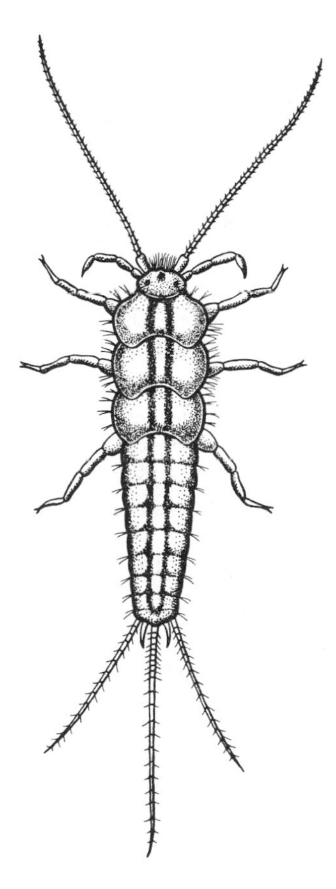

The silverfish. The body is covered with silvery scales, and whoever first named it may have been looking at one through his martini. Although very primitive, and believed to be descended from the earliest insects, before they had wings, silverfish still thrive in many homes. They are odorless and harmless, and serve to remind us of how ephemeral we are—and how ephemeral our books and records are, for silverfish live on the sizing of paper and the bindings of books. I keep my *American Scientist* in the attic for the silverfish to eat, although I have heard that they do equally well on *Playboy* or *The Watchtower*.

land rises. The increase in taxes forces many a well-meaning landowner to subdivide, and under the right of eminent domain other areas are gobbled up for superhighways. Land speculation is a primary source of easy money, and zoning lawyers thrive in the periphery of all our sprawling cities. Land worth a few thousand dollars an acre for housing may be worth much more for high-rise apartments or industrial plants.

Furthermore, all of Suburbia is pressed for tax money to provide schools for the growing hordes of children, and justly concerned with improving the quality as well as the quantity of schools. If it is a choice between preserving a woodlot as a park or sanctuary or rezoning the area for industry, one can be sure what decision will be made. In the town in which I live there was an excellent bog, inhabited by a host of plants and animals found nowhere else in the vicinity. The bog has now become the town dump—I use it myself—and when filled it will be zoned for industry. Certainly Lexington, Massachusetts, is a typical enough American town, more affluent than most, and therefore better able to afford the luxury of a sphagnum bog and a resident hermit thrush; but at the same time it is more concerned than most towns with the quality of its educational system—which is why I live there. Then, too, the bog breeds mosquitoes, and it is hard to persuade the residents that the preservation of several rare ferns, beetles, and birds is worth a few mosquitoes. If there is an easy solution to these problems, I should like to know it.

On the state and national scene, the picture at first seems brighter, since the number of state and national parks, forests, recreation areas, and wildlife refuges has grown steadily in recent years. In fact, the growth has by no means kept up with the pressure of a highly motorized public eager to spend their vacations in some new environment (though often enough carrying much of their old environment with them in their trailers). In 1966 there were 137 million visitors to areas of the National Park System—the equivalent of nearly 70 per cent of our population visiting areas that collectively total about 1 per cent of our land area. Human and automotive traffic is swamping the parks, and park rangers spend most of their time directing traffic, enforcing regulations, and giving an occasional fireside entertainment to the tourists. The Park Service has of necessity become more and more involved with tourism and less involved with the study of natural communities within their bounds: all the more tragically so since we are rapidly approaching the state wherein the *only* accessible remnants of the biosphere are in our parks and our state and national forests. A report by a committee of the National Academy of Sciences to the Secretary of the Interior in 1963 had this to say:

"It is inconceivable that property so unique and valuable as the national parks, used by such a large number of people, and regarded internationally as one of the finest examples of our national spirit should not be provided adequately with competent research scientists in natural history as elementary insurance for the preservation and best use of the parks."

The committee remarked that such research as was being conducted "has been marked by expediency rather than by long-term considerations." At the time the report was prepared, the National Park Service's research budget was about $28,000. It has since increased, but, in one commentator's words, is still "hardly enough to get Service scientists out of Washington's first taxi zone. . . . What has happened, it appears, is that the Park Service's policy makers seem to be succumbing to the recreation explosion at the expense of conservation" (Howard Simons, writing in *Bioscience*).

Only by basic ecological research in the parks can we begin to answer such questions as: Is the park large enough and of the right shape to constitute a proper habitat for elk, deer, trumpeter swans, or other animals? Are we justified in controlling certain bark beetles with chemicals or draining a marsh next to a campground? Where can a new road or campground be placed without causing undue erosion or destroying key elements in a natural community? What areas can safely be used for fishing, for educational purposes, or for recreational purposes such as motorboating and waterskiing, which can equally well be carried out in wholly "unnatural areas"? But the major need for basic research in the parks is simply because we have few other areas of reasonably inviolate nature—which is why I, for one, spend the greater part of my time in the field going from one state or national park or forest to another. The biologist would, of course, like to see many more preserves, not only in scenically attractive areas but in every possible natural situation. We have a tendency to make mountainous country into parks because it is of little value for agriculture or industry. I do not object to parks in mountains, but I regret that so little land is left unexploited in the valleys, plains, and seashores.

The parks are, of course, not only feeling the direct effects of the human population explosion but are also threatened by increasing demands for water, minerals, lumber, and other resources. The plan to flood part of Grand Canyon National Park has been shelved for the moment; but as of this writing, there are five new bills before Congress seeking authorization for one or more dams in the Grand Canyon. And we learn from the papers of the opposition of Indiana officials, from the governor down, to attempts to block filling and pollution adjacent to the proposed Indiana Dunes National Lakeshore: Bethlehem Steel is more important than a few sand dunes. As of this writing, many problems of the proposed Redwood National Park are still unsolved. In an interview with *Science*, Don Cave, a spokesman for the redwood lumber industry, is quoted as follows: "I say there's not going to be a Redwood National Park. We'll fight these bastards from hell to breakfast." Then, of course, there are the immortal remarks of Ronald Reagan during his campaign for governor of California: "A tree's a tree—how many more do you need to look at?" Reagan was, of course, elected by an overwhelming majority. Later he indicated a willingness to support such a park—but only if the National Forest Service gave up a 14,491-acre tract in exchange for land the government would acquire from lumber companies. A bill establishing a Redwood National Park passed the Senate on November 1, 1967, but as of this writing it has not reached the House. The bill includes a provision for a "swap" of lands along the lines suggested by Reagan, thus setting a dangerous precedent for the establishment of future parks.

Conservation has sometimes been called "ecology in action," but in fact is much more than that, calling into play as it does so many economic and political factors. Conservation foundations rely upon donations for their support, but according to law such donations can be considered tax-deductible only if the foundation uses no substantial part of its income to influence legislation. In contrast, industry, lumber interests, and others who profit from degradation of the environment are able to lobby freely, and money spent in this way is considered deductible business expense. Ironically, conservation organizations cannot even fight to have the Internal Revenue Code changed so that they may at least argue their case in legislatures. That the Code can be used as a weapon against conservationists was clearly shown in the case of the proposed flooding of part of the Grand Canyon; for the strong stand it took, the Sierra Club was informed by the Internal Revenue Service that contributions to the Club are no longer tax-deductible.

Those who believe that man cannot or should not live in a purely artificial environment not only need to be eternally vigilant on both the local and national scenes; they need also to be as tough-minded and politically knowledgeable as their opponents. The world picture is even more discouraging, for we must remember that more than half the people of the world receive less than we consider an adequate diet. One cannot expect the people who sleep on the streets of Calcutta to have developed an "ecological conscience." A person who does not know where his next meal is coming from is not likely to assume a broad view of the human endeavor (though in fact the Eastern religions have little of our Western arrogance toward nature). Nor can the uneducated be expected to look forward to tomorrow; the essence of education is its appreciation of the past and its concern for the future. The underdeveloped countries of the world want above all else to raise their standard of living, to share in the comforts produced by modern technology. In those countries with a population density already much greater than ours, and still increasing, there seems little prospect of ever giving much thought to conservation.

True, there are vast areas of the globe that are still relatively unaltered by man. But more and more the Sahara and other major deserts of the world are being mined and irrigated. We now read that Brazil is about to launch "an epic effort to achieve economic conquest of the Amazon basin." The Brazilians cannot be criticized (least of all by North Americans); at present less than 5 per cent of the soaring population of the nation lives in the million and a half square miles of the Amazon basin. It does indeed seem reasonable that all nations, all people should share alike in the bounty of the earth and in the rewards of modern technology. Now that the world has been rendered so small and so many persons who once accepted poverty and hunger as part of life have learned of Western affluence, any other view is not only inhumane but dangerous.

Realization that productivity is basically a problem in ecology and that it must be considered on a global scale led to the formation in 1963 of the "International Biological Program," which I find one of the more encouraging developments in recent years—although in fact it is merely an international coordinating body and has no operating budget. Persons concerned with the rapidly diminishing diversity of life on earth have heralded the IBP as a possible means of stemming the tide or at least of learning what we have before we lose it. At the Ninth Conference of Directors of Systematic Collections, held in Ann Arbor, Michigan, in April, 1965, the following resolution was passed:

"Whereas much native habitat is destroyed at the present time in all parts of the world, and replaced by agricultural plantings made necessary by the current population explosion, and whereas many species of animals and plants have extremely restricted areas of occurrence, and whereas many of these highly localized endemic species are threatened . . . , many others already having become extinct within recent decades, be it resolved that the International Biological Program add . . . to its program: (1) a fact finding survey and mapping of areas . . . which are most immediately threatened by current agricultural programs; (2) a faunal and floral survey of such areas, particularly in the tropics and subtropics, where indigenous and endemic species of animals and plants are most seriously threatened by extinction . . . (3) a sampling of threatened biota and preserving such samples in appropriate museums in this country or abroad."

As the anthroposphere continues to encroach upon the biosphere, the museums will more and more assume the role of guardians of the world's treasures. Just as one may now visit certain museums to see a stuffed passenger pigeon, a sad vestige of the great hordes that filled the skies not many decades ago, he may someday visit the museum to see other creatures that we now take for granted. Museums should not, however, be thought of primarily as places to look at curiosities, as "dead zoos."

They should be thought of as places for filing the products of biological evolution and of man's culture where they can be studied down through the years. More and more, natural history museums are becoming concerned not only with preserving dead specimens but also with preserving records of the living animal: nests, songs, behavior as recorded on film, and so forth. Furthermore, every major museum has one or more "field stations" and access to one or more "natural areas" where plants and animals can be studied as parts of their natural communities. Museums thus serve not only as centers for the identification of the two million or more kinds of organisms on earth but also as research centers on the role these organisms play in the biosphere.

One would hope that if we are indeed entering an "era of ecology," we would better appreciate the role of natural history museums as centers for the study of nature's diversity. I know of no institutions of comparable importance that are so grossly undersupported. In my own museum we have but two insect curators and two underpaid assistants for a collection said to contain about five million insect specimens. Even the Smithsonian Institution, for all the publicity it has been receiving in the past few years, is grossly understaffed in entomology and doubtless in other fields. The United States Department of Agriculture maintains a team of taxonomists (located at the Smithsonian) charged with the task of identifying the fantastic number of insects submitted for identification—usually in connection with research or control of pest species. Yet in fact the number of specialists in the Division of Insect Identification of the Agricultural Research Service actually declined from twenty-five to twenty-one during the period 1932–1962, and during the same time the amount of time each specialist spent on basic research declined from an average of 63 per cent to about 25 per cent. This in spite of the fact that many of our insect problems have become increasingly acute, and we still lack even a preliminary understanding of some groups of insects of vast size—the parasitic wasps, for example. Will our current preoccupation with the human environment one day result in a "new deal" for our museums, or will we continue to try to prognosticate the future from a crystal ball that is murky to the point of being altogether opaque?

The United States Committee for the International Biological Program has recently issued a brochure describing the programs under way. One is titled "aerobiology," and concerns the patterns of dispersion of pollen grains, fungus spores, small insects, and other objects by air currents—a matter with a good deal of practical importance as well as of value in helping to understand the distribution and evolution of plants and animals generally. Another is titled "analysis of ecosystems," and consists of in-depth studies of selected natural communities from the point of view of levels of productivity under various natural and disturbed conditions. A third proposes to intensify biological study of the Hawaiian Islands while there are still a few niches not fully occupied by man. One could easily think of a great number of problems equally worthy of support. Did I say support? Each of the projects I mentioned is marked "funding to be developed"!

If we do undertake the biological surveys envisaged by the IBP, on the assumption that the knowledge gained will help us to utilize the abundance of life on earth while still preserving much of its diversity, we shall be up against the hard fact that whatever terrestrial community we study will be found to contain great numbers of insects, and a high percentage of these insects (especially in the tropics) will be unknown to science, or at least known in only the most superficial way. Also, the complexity of each community is such that we cannot, a priori, designate any particular species as unworthy of study. A midge dancing over a patch of moss on Baffin Island may prove to be just as important a key to understanding as a rat in the ghettos, an elephant trampling the African brush. This being the case, I hope you will forgive this display of evangelism on the part of a person who usually works quietly in the midst of a collection of millions of dead insects—with no one to sound off to but them.

Whether he likes it or not, man must concern himself with those remarkably diverse and prolific coinhabitants of the earth, the insects. As William D. Peck put it in 1819:

"These diligent and faithful servants of nature, as Linnaeus calls them, are perpetually engaged in destroying all that is dead, and of checking the increase of all that is living in the vegetable world. In the execution of the task assigned them, they often frustrate the designs and subvert the arrangements of man, thus constraining him to attend to objects which are generally deemed beneath his notice, and obliging him to feel how effective is the smallest instrument in the hand of Omnipotence."

It is ironic that these words are as true today as they were a century and a half ago. In spite of our technological advances, the insects still "subvert our arrangements," and in the course of combating them we add a variety of poisons to an environment already flooded with our wastes. We sometimes speak of bypassing the living world by developing wholly synthetic foods. We seem to forget that chlorophyll, the green substance of plants, remains and will remain the essential link for converting solar energy to human energy. This means that we shall have to continue to be concerned with insects and all their complex interrelationships with plants: the great numbers that serve as pollinators and without which so many of our fruits would not set; the parasitic wasps and flies, which act as a brake against the overwhelming increase of plant-feeders; the many kinds that assist in decomposing dead plant and animal material and returning it to the soil; and so forth and so on. We have, indeed, learned a great many things since Peck's day, but we are still continually surprised by the ability of insects to resist our potent chemicals, to burst forth in great numbers in unsuspected places.

It would seem especially pertinent to delve into the complexities of population phenomena at a time when we ourselves are embarked on an explosion in numbers without precedent for our species. No other group of organisms provides so many examples of sudden gross increase in numbers as do the insects: the appearance of "snow fleas" in incredible numbers where none were noted before; the swarms of plague locusts; the build-up of plant lice in shade trees to the extent that they drip honeydew all over our latest status symbols from Detroit; the periodic explosion of a new "pest" species. In these cases (and many others) the explosion sooner or later ends, and with tragic overtones. Are there ways of maintaining an environment suitable for continuing a high population, and if so what do they imply in the way of resources and relationships with other species? The study of natural outbreaks of insect species, and the reproduction of population oscillations under laboratory conditions, may provide us with needed insights into our own population problems.

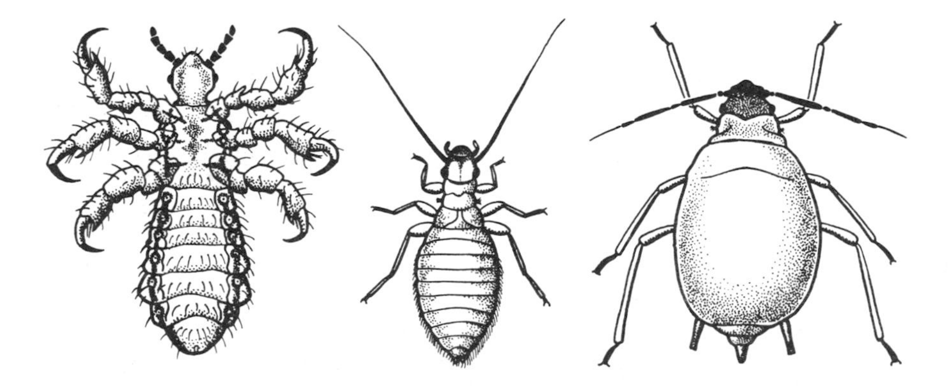

It's a lousy world. The body louse (*left*) has become resistant to DDT, and the plant louse or aphid (*right*) thrives in Suburbia's shade trees and dribbles on our shiny new automobiles. The book louse (*center*) has provided a decorative motif for this book; doubtless he is thriving in the midst of our current information explosion, though he is too small to be noticed in our busy world. The book lice in my study are, however, appreciated. Without them I might easily have been diverted to the trivialities of humanity's grand affairs and never finished this tribute to the importance of the minuscule.

If present trends continue, it is certain that the question "Is nature necessary?" will soon become academic. Nature will become a luxury we cannot afford—the anthroposphere, with its attendant management to maintain the "balance of man," will have covered virtually all the land surface, and doubtless some of the oceans. Can we soon reach a decision on how many people the earth can support in the light of human ideals as to how life should be lived?

This is clearly the major problem of our time, beside which all other problems pale into insignificance (unless one considers war to be a separate problem, which I do not). Despite all the recent discussion and the excellent progress in developing new methods of contraception, human numbers continue to explode, and the attendant revolution of the environment continues at an accelerated pace. It will do no good to shake our fingers at the bulldozers and the pest-control operators. What is needed is a counterrevolution of major proportions. It is the biologists who must lead such a counterrevolution, since in fact man's problems are biological, and so, in large part, are their solutions. Biology, after all, is the study of living things, and as such surely includes man, his mores, and his artifacts. That this is not often recognized is the source of much of our incapacity to meet our growing crises. We know, for example, that a continuous growth rate in the numbers of any species, if continued indefinitely, will result in the total exhaustion of resources; and man, for all his cleverness, cannot make food out of rocks. A human growth rate of zero is therefore desirable now, soon, or perhaps was desirable a few years or decades ago. That we scarcely dare mention such a thing, but prattle about "family planning" and the like, is a reflection of our willingness to be guided not by biology but by taboos and dictates inherited from prescientific eras (at the same time accepting, of course, the greatly reduced death rates brought about by scientific discoveries). In a recent article in *Science*, Kingsley Davis presents some startling figures about "family planning" in several Latin American cities. Surveys showed that women wished to have, on the average, 3.4 children; but the actual births per woman averaged 3.7. Thus, even if these women used contraceptives with complete efficiency, "they would still have enough babies to expand city populations senselessly."

Even the most optimistic demographers think it unlikely that we can stop our population explosion short of two more doublings, that is, at about 13 billion people. What does this mean in terms of the quality of human life? We know that many animals—deer and meadow mice, for example—when severely crowded show enlargement of their adrenal glands and severe symptoms of stress; they may fail to reproduce or even die off in spite of abundant food. Recent studies show that when certain male cockroaches are caged together they fight repeatedly, and those that are consistent losers often enter a semiparalyzed state and finally die. Wild rats fight extensively among themselves, but laboratory rats, stocks of which have been under domestication for a hundred years or more, are much tamer and more tractable; they have smaller adrenal glands and smaller brains, but they have larger sex organs and spend much more time in sexual activity than do their wild relatives. Perhaps civilized man is more like the laboratory rat, and is evolving a type that is almost infinitely compressible so long as he is able to eat and reproduce.

In Calcutta, people live at a density of 102,000 per square mile, yet life goes on. Presumably many parts of the world will approach and eventually surpass that density unless something radical is done. For of course it is not simply a matter of research; it is also a question of our philosophy of life. Do we wish our children to live in a world that is above all well regulated and filled with social contacts; or is there something to be said for a world that is filled with variety, mobility, a measure of freedom and challenge, and an opportunity for solitude? We can be quite sure that our children will live in a much more restricted and thoroughly regulated world, just as we in our generation can look back to the time when there were no traffic lights and parking meters. Where we want to draw the line on the expansion of human numbers and artifacts is our decision, and the decision will depend upon what we think life ought to be. Is the urbanization of the world what we really want? Are more and larger cities and suburbs really desirable, and do we wish to devote all intercity space to feeding and entertaining the urbanites? The frightening riots in the ghettos should cause us to ask whether we are on the right course. We hear, too, that major crime has doubled in the United States in only three decades, and is growing proportionally faster than our population. Juvenile delinquency has reached the point where one in six youths is charged with a crime other than the common one of auto theft. Five and a half million Americans are mentally ill. Even the prosperous suburbanite, oblivious as he is to most of the world's ills, has cause to ponder when his car is caught in a continuous traffic jam, when his new cruiser is set in waters foul with the wastes of our exploding population.

There are some who believe it is already too late for the world to avoid widespread famine. In a widely publicized speech given at the University of Texas in November, 1967, Stanford biologist Paul Ehrlich remarked that "it is shockingly apparent that the battle to feed humanity will end in a rout. . . ." He suggests that the first step "must be to convince everybody to think of the earth as a space-ship that can carry only so much cargo. When we have determined the proper size of the crew, then we can design the environment to suit."

It is customary to blame our population problems on the Church and on the male "cult of virility," but there are other factors, perhaps more important in the long run: the fear of being overrun by another nation or another race; above all, the fear of economic stagnation. To the American economy there is no better news than that our population will double in about fifty years: think of it, we shall need twice as many automobiles, roads, factories, dams, airports, and so forth! Our Gross National Product (GNP) is expected to grow to nearly one trillion dollars by 1975, and you may be sure that any administration that does not maintain our yearly 3 to 4 per cent increase is not likely to be in office long. Somehow the words growth and progress have become confused. Uncontrolled growth in a tissue is called cancer, and calls for surgery or some other desperate measure; in an animal population it is called an imbalance or outbreak, and calls for pesticides or a longer hunting season. Is man above all this? Can his economy mushroom indefinitely without draining away all that has nourished it? In the final analysis, is not progress to be measured not in the height of our piles of dollars but in the depth of our understanding of the meaning of life? The idea that the GNP is not the measure of all things is shocking to most Americans. A nation that is used to having its landscapes partially obscured by billboards and its most serious news programs interrupted by jingles on behalf of some ridiculous luxury is perhaps beyond saving. But there is hope to be found in the rebellion of youth against "the establishment," in the insidious spread of that subversive attitude called "an ecological conscience." Tomorrow's world is apt to be very different, one way or another.

Those who decry what they call "the mania of eternal growth" nevertheless still have a lot of convincing to do. One major difficulty is that biologists and conservationists love to discourse among themselves, where they will have a sympathetic hearing without rebuffs from the hard worlds of business and politics. Some of the best essays in this field have been in journals such as *Bioscience* and *National Parks Magazine*, which rarely invade Wall Street or the halls of Congress. To be sure, Western man has become distinctly more environment-oriented in the past decade. Yet it is difficult for him to grasp the great many subtle and often apparently trivial factors that make up what we glibly call "the environment." Even the ecologists often lack the data and experience to predict the consequences when the temperature of a stream is increased several degrees by the effluent from an atomic energy plant, when a tract of desert is irrigated or forest cleared, when a plant or animal is eliminated or introduced into a new area. Because of the many complex variables involved, ecology cannot put forth the simple message and promise of quick breakthroughs needed to attract widespread interest and support. Those concerned with preservation of the environment are often cast in the role of starry-eyed sentimentalists too concerned with fuzzy Utopias and prophecies of doom to grasp the world of reality. Yet it can be argued, as I would argue, that the thin film of life that surrounds the rocky core of the earth is a far more delicate and precious thing than we have yet been able to grasp. Its most wonderful manifestation, the human species, cannot disavow or disregard the biosphere without withering on the vine that engendered it.

There are few who would deny the need to understand more fully the roots of human behavior. The controversies aroused by Konrad Lorenz' book *On Aggression* and Desmond Morris' *The Naked Ape* reveal how little we know about this subject. This knowledge can come only from a far broader knowledge of animal behavior generally than we now possess. And since behavior evolves in and is closely conditioned by the community in which an animal occurs, it is fully understandable only in an ecological context. It is this field—behavior and environment, or ethology and ecology—that is truly the science of tomorrow.

The mere accumulation of facts is a fascinating business—living organisms being what they are—and it is one in which every person with a background in biology can participate. (And all things considered, should not everyone be trained in biology?) Right now we are engaged in a losing battle: the rate of extinction of species exceeds our rate of learning about them. That we should let this happen is not a tribute to the intelligence of which we as a species are so proud. Of course, we need more than facts; as our knowledge grows, we need to seek generalities, the unity to be found in multiplicity. A scientific hypothesis is very much like a successful painting: something unsuspected and beautiful is revealed. This, too, is within the province of every educated person with the facts at his disposal. The best insights sometimes come from persons who are not specialists, who are not too lost among the trees to see the forest.

The study of living things is not only something to pursue because it is important and grossly underdeveloped. It is also a great deal of fun. To a person attuned to smaller creatures (and man, sitting on top of the pyramid of numbers, must be so attuned) there is no corner of nature not full of excitement, not rich in unsolved problems. Every home should, of course, have a microscope, for only a small fraction of the living world can be appreciated with the naked eye. Insects provide wonderful subjects for study from every point of view, and it is hard to study them for very long without coming up with a new fact or concept. Insects have taught us a great deal about animal diversity and evolution; about genetics, growth, and the chemistry of life; about the interrelationships of living things, the significance of animal numbers. They will teach us a great deal more, and entertain us in the process.

In these pages I have sketched the lives of a few of the insects I admire. But I admit to being grossly ignorant of algae and earthworms and a great many other forms of life. The fact that there are literally millions of species of plants and animals on this little-known planet of ours is to me overwhelmingly exciting—for each species has a slightly different role to play in a drama that we can now reconstruct (in a general way) from its beginnings—but can scarcely predict beyond this moment. Furthermore, every character in the play in some way influences or has influenced every other in a gigantic web of interrelationships far beyond our comprehension. The role and destiny of man are not fully clear, but it is unimaginable that he should someday perform alone on the stage—or that the play should go on without him.

The earth is a good place to live. We shall appreciate it more and more as we explore the moon and the planets. If man shall ever have another home, it is presently unimaginable. We had better learn to respect the little-known planet beneath our feet. In the words of that most quotable of moderns, the late Adlai Stevenson:

"We travel together, passengers on a little space ship, dependent on its vulnerable reserves of air and soil; all committed for our safety to its security and peace; preserved from annihilation only by the care, the work, and, I will say, the love we give our fragile craft."

Notes on Classification

Chapter 2 Cities in the Soil: The World of Springtails

The springtails make up the order Collembola (coh-lem′-boh-la), based on the Greek words meaning "glue-peg," as explained in the text. There are reported to be about 5,000 species in the world as a whole, about 600 in North America.

Since most springtails have no common names, I had to use a number of scientific names in this chapter. Sminthurides (smin-thur-eye′-deez) is Greek for mouselike; Dicyrtomina (die-sir-toh-meen′-uh) is from the Greek words meaning "a double humpback"; Hypogastrura (hi-poe-gas-tru′-ruh) is from the Greek for "tail beneath the belly"; and Isotoma (eye-so-tome′-uh) is from Greek words meaning "equal divisions," referring to the even segmentation of the body. The species name nivicola is Latin for "living on snow"; the species names aquaticus, minuta, and sepulcralis mean just what they seem to mean. Like most scientific names, these are attractive, appropriate, and easy to pronounce—and they do not change from country to country. We need to nurture the few international languages we have.

Chapter 3 The Intellectual and Emotional World of the Cockroach

The cockroaches belong to the order Dictyoptera (dick-tee-op′-terr-uh), based on the Greek words meaning "net-wing," with reference to the delicate network of supporting veins on the front wings. Two other kinds of insects closely related to cockroaches also belong to this order: the mantids and the termites. The average termite may not look much like a roach, and many books will tell you that the termite belongs to a different order, the Isoptera. But the fact is that every line of evidence suggests that termites are "socialized" cockroaches, and two peculiar animals still survive that connect these two groups. One is a large and very roachlike termite from Australia; the other is a termitelike roach that lives in rotten logs in the southern Appalachians. The latter is semisocial, and even has one-celled organisms in its digestive tract that help in digesting wood—just as the termites do. An attractive and readily available introduction to termites, by the way, is S. H. Skaife's book, *Dwellers in Darkness* (Doubleday Anchor Books, 1961, paperback, 180 pages).

The cockroaches as a whole make up a group (suborder) known as the Blattaria, which is based on the Latin word for roach, *blatta*. Blatta is also the generic name for the Oriental roach, Blatta orientalis, while its diminutive form provides the generic name for the German roach, Blattella germanica. In the course of this chapter, I used several scientific names of species lacking common names. Diploptera punctata (dip-lop′-terr-uh punk-tate′-uh), literally "double-wing covered with small holes," is one of the most beetlelike of roaches. Nauphoeta cinerea (naw-fee′-tuh sigh-near′-ee-uh) is a name compounded of the Greek word *naus*, a ship, plus *phoitetes*, one who comes and goes, to which is added the Latin adjective *cinerea*, ashy-colored (that is, an ash-colored roach that travels on ships, which it sometimes does). The wonderful name Gromphadorhina portentosa (grom-fad′-o-rine-uh pore-ten-tose′-uh) is compounded of the Greek words *gromphas*, an old sow, plus *rhinos*, snout; the word "portentosa" means just what it sounds like, portentous or ominous. Blaberus (blab′-er-us) is from the Greek *blaberos*, meaning noxious; while the species name "craniifer" (crane′-ee-if-er) is from the Latin *cranium*, a skull, plus the verb meaning to bear, *ferre*.

Chapter 4 Water Lizards and Aerial Dragons

Dragonflies and damselflies belong to the order Odonata (oh-don-ate′-uh), based on the Greek word for tooth, *odontos*. This is apparently an allusion to the powerful teeth on the mandibles of the adult and the lower lip of the larva. Libellula (lie-bell′-you-luh), as we have already mentioned, is a generic name based on the Latin word *libella*, a balance. The name of the green darner, Anax junius (ay′-nacks jew′-nee-us), is from the Greek word for monarch, *anax*, plus the Latin word for the month of June, *junius*: the monarch of June. Plathemis (plath′-em-is) is presumably a contraction of the Greek words *platys* + *themis*, literally "broad justice," while the species name lydia is simply a girl's name. Argia (ar′-gee-uh) is probably based on the Greek word *argos*, meaning "bright," while the species name apicalis is based on the Latin stem that gave us our words "apex" and "apical": a reference to the bright spot on the apex of the abdomen of these attractive damselflies.

Chapter 5 The Cricket as Poet and Pugilist

The crickets, katydids, and grasshoppers belong to the order Orthoptera (or-thop′-terr-uh), based on the Greek for "straight-wings," with reference to the relatively straight and narrow wings of many of these insects. The Latin for "cricket," *gryllus* (grill′-us), has provided the generic name for the field crickets, and forms the basis of the family name of the crickets, Gryllidae (grill′-id-ee). The name of the lesser field crickets mentioned by Snodgrass, Nemobius (nem-oh′-bee-us), is based on the two Greek words *nemos* (pasture) plus *bios* (life). Those most elegant of all songsters, the tree crickets, belong to the genus Oecanthus (ee-can′-thus), which is based on the Greek words *oecos* (home) plus *anthos* (flower). The name Cyphoderris (sie-foe-der′-iss) is from the Greek *cyphus*, humpbacked, plus *derris*, a leather cloak. The name of the small, voiceless cricket that lives with ants is Myrmecophila (mer-meh-kof′-ill-uh), based on the Greek for "ant-lover" (*myrmex* + *philia*).

The katydids are related to the crickets but comprise a separate family, the Tettigoniidae (tet-ee-gohn-ee′-id-ee), based on the Greek for a small singing insect, *tettigonion*. The grasshoppers (which when migratory are called locusts) constitute a second major group of Orthoptera, having much shorter antennae and ovipositors than crickets and katydids, as well as very different sound production and hearing organs. Grasshoppers belong to the family Acrididae (ak-rid′-id-ee), based on the Greek for "grasshopper," *acris*. They are discussed at length in Chapter 10.

Chapter 6 In Defense of Magic: The Story of Fireflies

Fireflies, like other beetles, belong to the order Coleoptera (coal-ee-op′-terr-uh), a name based on Greek words meaning "sheath-wings," with reference to the fact that the front wings form more or less hardened covers for the hind wings. The true fireflies belong to the family Lampyridae (lam-pee′-rid-ee), from the Greek word for the glowworm, which is in turn based on the words *lampein*, to shine, + *ouron*, tail. The generic name Photuris (foe-tour′-iss) has much the same derivation, *photos* being the Greek word for "light," and *ouron*, again, for "tail." Photinus (foe-tine′-us), the name of the smaller fireflies, can appropriately be translated "a little light," the suffix "inus" being a diminutive. Photinus pyralis gets its species name from the Greek word for "fire," *pyr:* literally, this species is a fiery little light.

I also mentioned Pyrophorus (pie-rahf′-o-russ), the insect with paired greenish lights just behind its head and an orange light on its abdomen; this name is also based on the Greek word for "light," *pyr*, + *pherein*, to bear. This insect belongs to the family of click beetles, Elateridae (ee-lah-terr′-id-ee), the Greek word *elater* being the word for a hurler, with reference to the way in which these insects are able to propel themselves by "clicking" a special snap on their underside.

There are so many beetles that volumes could be written about them (although a good book on the biology of beetles has yet to be written). There are at least 210,000 species, and it is probable that when all have been discovered and described there will prove to be a great many more than that. When the British biologist J. B. S. Haldane was asked what one could conclude about the nature of the Creator from a study of His creations, he is said to have replied: "an inordinate fondness for beetles."

In this chapter I also mentioned certain luminous gnats, including the New Zealand glowworm. These are true flies, of the order Diptera, a group we look at from various other points of view in Chapter 8.

Chapter 7 Interlude in the Elysian Meadows: Butterflies

Butterflies belong to the order Lepidoptera (lep-id-op′-terr-uh), which is Greek for "scale-winged." The vast majority of Lepidoptera (and there are more than 100,000 species) are night-flying moths that only rarely come to the attention of nonspecialists. Butterflies can generally be told by their slender, knobbed antennae; they make up only one superfamily of this large group, the Papilionoidea (pa-pill-ee-on-oy′-dee-uh). This name is formed from the generic name of the swallowtails, Papilio (pa-pill′-ee-o), which is the Latin word for "butterfly." The species name of Papilio dardanus comes from Dardanus, the mythical ancestor of the Trojans, son of Jupiter by Electra, daughter of Atlas. Heliconius erato (hell-ee-cone′-ee-us er-ah′-toe) is also a name based on Greek mythology. Erato was the muse of lyric poetry, while Heliconius means "pertaining to Helicon," a mountain in Greece that was sacred to the muses. The generic name Ornithoptera (or-nith-op′-terr-uh) is simply Greek for "bird-wing." Hypolimnas misippus (hi-poe-lim′-nahs miss′-ip-us) is also from the Greek, Misippus being a mythological character, hypolimnas compounded from words meaning "below the marsh."

Chapter 8 Paean to a Volant Voluptuary: The Fly

The true flies comprise the order Diptera (dîp′-terr-uh), which is simply Greek for "two-winged." The Greek word for fly was *myia,* which forms the stem of such words as "myiology" (the study of flies) and "myiasis" (infection by flies). The Latin word for fly, *musca,* has come down to us as the generic name of the housefly, Musca domestica. Some of the genera of flies are so well known as to need little explanation: Drosophila (dro-sof′-ill-uh), Greek for "lover of dew"; Anopheles (an-ah′-fell-eez), the malaria mosquito, Greek for "hurtful"; Aedes (aye-ee′-deez), the yellow-fever mosquito, Greek for "disagreeable." Just to prove that all flies' names are not that simple, let me mention Monochaetoscinella anonyma: a fly so small it would take at least 100 to cover up its name. One of my favorites, though, is a fly parasitic on other insects, and called Gremlinotrophus derisus, which I translate to mean "the laughing Gremlin-eater." It was given its name by H. J. Reinhard, of Texas A.&M., a man who has devoted a long life to enjoying flies and helping to diminish our ignorance about them.

Chapter 9 Bedbugs, Cone-nosed Bugs, and Other Cuddly Animals

The true bugs belong to the order Hemiptera (he-mip′-terr-uh), which is Greek for "half-wings." They are so named because the basal part of the front wings is thickened, the outer part thin and transparent. Actually, the most distinctive feature of the order is the sucking beak, made up of four long piercing stylets that at rest are enclosed in a sheath. All Hemiptera feed on fluids, either plant sap or the blood of animals.

The bedbug and its immediate relatives make up the family Cimicidae (sigh-miss′-id-ee), based on the Latin word for bug, *Cimex* (sigh′-meks). The cone-nosed bugs comprise the family Reduviidae, based on the generic name of the masked bedbug hunter, Reduvius (red-you′-vee-us) (Latin for a "sore on the finger"). The name Rhodnius (rod′-nee-us) is probably based on the Greek word for "slender," *rhadinos,* while the species name, *prolixus* (pro-licks′-us), is the Latin word for "long." In the last part of the chapter we also met the plant-feeding bug Pyrrhocoris apterus (pie-roe-core′-iss ap′-terr-us), based on the Greek words for "wingless" (*apterus*) "flame-colored" (*pyrrhos*) "bug" (*coris*). This bug belongs to the family Pyrrhocoridae, which is somewhat distantly related to the two families of bloodsuckers.

We also met some of the silk moths, which of course belong to the Lepidoptera, an order that we looked at from various other angles in Chapter 7.

Chapter 10 Year of the Locust

As already mentioned, the grasshoppers and locusts are classified in the same group, the family Acrididae (a-krid′-id-ee), based on the Greek word for "grasshopper," *acris.* The Latin word for grasshopper, *locusta,* has provided the generic name for the migratory locust, Locusta migratoria (lo-cuss′-tuh my-gra-tor′-ee-uh). The desert locust is called Schistocerca gregaria (shis-toe-sir′-kuh greg-air′-ee-uh), based on the Greek words *schistos* (divided) + *cercos* (tail), with reference to a notch at the tail end of the male; *gregaria,* of course, is simply Latin for gregarious. The Rocky Mountain locust has the interesting scientific name Melanoplus spretus (mel-an′-o-plus spree′-tus), from the Greek *melanos* + *oplon* and the Latin adjective *spretus:* literally, "a despised creature clad in dark armor"—a thoroughly suitable name. All these insects belong to the same order as the crickets of Chapter 5, the Orthoptera.

The cicadas, which are sometimes erroneously called locusts in the United States, belong to the order Hemiptera, other members of which we discussed in the preceding chapter.

Chapter 11 Parasitic Wasps, and How They Made Peyton Place Possible

The wasps make up the greater part of the order Hymenoptera (hi-men-op′-terr-uh), a word derived from the Greek for "membranous-wings." This order also includes the so-called "sawflies," which might more properly be called "wood wasps" and "leaf wasps" (as they often are in other parts of the world). It also includes two groups of remarkable insects that are believed to have evolved from wasplike ancestors: the ants and the bees. For an outline of the parasitic and predatory wasps, with many illustrations and much information on life histories and behavior, see C. P. Clausen's excellent book *Entomophagous* (i.e., insect-eating) *Insects.*

There are so many wasps and they are so little known to the average nonspecialist that hardly any of them have common names. Consequently, I have had to use many scientific names in this chapter, and I apologize. The following is a list of the names used, in outline form:

Ichneumonidae (ik-new-mahn′-id-ee): based on the Greek word *ichneumon*, a tracker. This is an enormous family, and includes most of the larger parasitic wasps. The two parasites of the cecropia belong to the genus Gambrus (gam′-brus) (Greek *gambras*, related) and to the genus Mastrus (mass′-trus) (of unknown origin).

Braconidae (bra-kon′-id-ee): a name of unknown origin. This is a large family related to the preceding. We mentioned one genus, Apanteles (uh-pan′-tell-eez), which is derived from the Greek *a* (not) + *panteles* (complete), with reference to the structural reductions in these small insects.

Chalcidoidea (kal-sid-oy′-dee-uh): based on the Latin word *chalcis*, a metallic-colored wasp, which is in turn based on the Greek word for copper, *chalcos*. This is a superfamily containing numerous families of mostly very small wasps. Trichogramma (trick-o-gram′-uh) is a generic name based on the Greek words *trichos* (hair) + *gramma* (a mark) ("a hairy little spot"). We mentioned two species: evanescens, which is a Latin adjective meaning just what it seems to mean—"now here, now gone"; and semblidis, which may be based on the Greek *symbletos*, meaning "comparable." The tongue-twister Ooencyrtus submetallicus (oh-oh-en-sir′-tus sub-meh-tal′-ic-us) is from the Greek meaning "a humpbacked egg parasite" + the Latin word meaning "somewhat metallic."

The aquatic egg parasite Caraphractus cinctus (care-uh-frak′-tus sink′-tus) is named from the Greek words *cara* (head) + *phractos* (partitioned), along with the Latin adjective *cinctus* (girdled)—all with reference to certain structural features. The generic name Litomastix (lie-toe-mass′-ticks) is from the Greek *litos* (simple) + *mastix* (whip). Coccophagus (koks-ahf′-o-gus) is Greek for "eater of scale insects," while Aphytis (ay-fie′-tiss) is of doubtful origin, perhaps from *a* (without) + *phyton* (plant). The three genera associated with cecropia are as follows: Dibrachys (die-brak′-eez) (Greek *dis*, twice + *brachys*, short), Pleurotropis (plu-roe-trow′-pis) (Greek *pleura*, side + *tropis*, keel), and Dimmockia (dim-ock′-ee-uh) (named after George Dimmock, an entomologist who lived in Springfield, Massachusetts). One group we mentioned by their family name was the Eucharitidae (you-kar-it′-id-ee), odd little ant parasites that owe their name to the Greek words *eu* (true) + *charitos* (loveliness).

Scelionidae (sell-ee-on′-id-ee): a family of egg parasites not closely related to the Chalcidoidea, deriving its name from the Latin word for "scoundrel," *scelio*. This group includes Wilson's Asolcus basalis (ay-soul′-kus bay-sal′-iss); *basalis* is a Latin adjective meaning "basal," while Asolcus is a name of unknown origin.

Trigonalidae (trig-oh-nall′-id-ee): from the Greek word *trigonos*, triangular, with reference to a triangular cell in the wing.

Dryinidae (dry-in′-id-ee): probably the diminutive of *Dryas*, a nymph.

Bethylidae (beth-ill′-id-ee; or sometimes beth-ile′-id-ee): apparently a combination of letters put together by the French biologist Latreille because he liked their sound. Perhaps, after hours of browsing through his Greek and Latin dictionaries, he could find no words that did justice to these wonderful little wasps.

For Further Reading

Chapter 1 **The Universe as Seen from a Suburban Porch**

Here are a few good, recent books on insects that present much general information. All are well illustrated, and the first two and fourth have many excellent photographs in color.

Farb, Peter, and the editors of *Life*. 1962. *The Insects*. Life Nature Library; Time, Inc., New York. 192 pp.
Klots, A. B., and E. B. Klots. 1959. *Living Insects of the World*. Doubleday & Company, New York. 304 pp.
Lanham, Url. 1964. *The Insects*. Columbia University Press, New York. 292 pp.
Newman, L. H. 1966. *Man and Insects: Insect Allies and Enemies*. Nature and Science Library; Natural History Press, New York. 252 pp.
Oldroyd, Harold. 1962. *Insects and Their World*. Phoenix Books, University of Chicago Press (paperback). 139 pp.

Chapter 2 **Cities in the Soil: The World of Springtails**

Christiansen, K. D. 1964. *The Bionomics of Collembola*. Annual Review of Entomology, Vol. 9, pp. 147–148.
Farb, Peter. 1962. *Living Earth*. Pyramid Publications, New York. 160 pp.
Kevan, D. K. McE. 1962. *Soil Animals*. Philosophical Library, New York. 237 pp.
Mills, H. B. 1934. *A Monograph of the Collembola of Iowa*. Collegiate Press, Ames, Iowa.

Chapter 3 **The Intellectual and Emotional World of the Cockroach**

McKittrick, F. A. 1964. *Evolutionary Studies of Cockroaches*. Memoir, Cornell University Agricultural Experiment Station, No. 389. 197 pp.

McKittrick, F. A., T. Eisner, and H. E. Evans. 1961. *Mechanics of Species Survival.* Natural History, Vol. 70, pp. 46–51.

Roth, L. M., and E. R. Willis. 1954. *The Reproduction of Cockroaches.* Smithsonian Miscellaneous Collections, Vol. 122, No. 12. 49 pp.

——. 1957. *The Medical and Veterinary Importance of Cockroaches.* Smithsonian Miscellaneous Collections, Vol. 134, No. 10. 147 pp.

Chapter 4 Water Lizards and Aerial Dragons

Corbet, P. S. 1963. *A Biology of Dragonflies.* Quadrangle Books, Chicago. 247 pp.

Corbet, P. S., Cynthia Longfield, and N. W. Moore, 1960. *Dragonflies.* The New Naturalist; Collins, London. 260 pp.

Needham, J. G., and M. J. Westfall, Jr. 1955. *A Manual of the Dragonflies of North America.* University of California Press, Berkeley and Los Angeles. 615 pp.

Walker, E. M. 1953, 1958. *The Odonata of Canada and Alaska.* Vol. I: *Damselflies;* 292 pp. Vol. II: *Dragonflies;* 318 pp. University of Toronto Press, Toronto, Canada.

Chapter 5 The Cricket as Poet and Pugilist

Alexander, R. D. 1961. *Aggressiveness, Territoriality, and Sexual Behavior in Field Crickets* (Orthoptera: Gryllidae). Behaviour, 17: 130–223.

——. 1966. *The Evolution of Cricket Chirps.* Natural History, 75: 26–31.

Haskell, P. T. 1961. *Insect Sounds.* Quadrangle Books, Chicago. 189 pp.

Laufer, Berthold. 1927. *Insect-Musicians and Cricket Champions of China.* Field Museum of Natural History, Chicago; Anthropology Leaflet 22. 27 pp.

Chapter 6 In Defense of Magic: The Story of Fireflies

Buck, John B., and Elisabeth Buck. 1966. *Biology of Synchronous Flashing of Fireflies.* Nature, 211: 562–564.

Harvey, E. Newton. 1957. *A History of Luminescence from the Earliest Times until 1900.* Memoirs of the American Philosophical Society, 44: 1–692.

Lloyd, James E. 1966. *Studies on the Flash Communication System in Photinus Fireflies.* Miscellaneous Publications, Museum of Zoology, University of Michigan, No. 130. 95 pp.

McElroy, W. D., and H. H. Seliger. 1962. *Biological Luminescence.* Scientific American, 207 (6): 76–89.

Chapter 7 Interlude in the Elysian Meadows: Butterflies

Brower, Lincoln, and others. 1963. *Mimicry; a symposium.* Proceedings of the XVI International Congress of Zoology, Washington, Vol. 4, pp. 145–186.

Carpenter, G. D. H., and E. B. Ford. 1933. *Mimicry.* Methuen, London. 134 pp.

Cott, Hugh B. 1940. *Adaptive Coloration in Animals.* Methuen, London. 508

pp., 48 plates. (Reprinted with minor changes in 1957; reissued in 1964 by Dover Publications, New York).

Ford, E. B. 1945. *Butterflies.* The New Naturalist; Collins, London. 368 pp.

Klots, A. B. 1951. *A Field Guide to the Butterflies of North America, East of the Great Plains.* Houghton Mifflin, Boston. 349 pp.

Chapter 8 Paean to a Volant Voluptuary: The Fly

Bates, Marston. 1949. *The Natural History of Mosquitoes.* The Macmillan Company, New York. 379 pp.

Dethier, V. G. 1962. *To Know a Fly.* Holden-Day, Inc., San Francisco. 119 pp.

Knipling, E. F. 1960. *The Eradication of the Screw-Worm Fly.* Scientific American, Vol. 203, pp. 54–61.

Oldroyd, Harold. 1965. *The Natural History of Flies.* W. W. Norton & Co., New York. 324 pp.

Roth, M., L. M. Roth, and T. E. Eisner. 1966. *The Allure of the Female Mosquito.* Natural History, Vol. 75, pp. 26–31.

Chapter 9 Bedbugs, Cone-nosed Bugs, and Other Cuddly Animals

Usinger, Robert L. 1966. Monograph of Cimicidae. Thomas Say Foundation, Vol. 7. Entomological Society of America, College Park, Md. 585 pp.

Wigglesworth, V. B. 1959. *The Control of Growth and Form.* Cornell University Press, Ithaca, N.Y. 140 pp.

———. 1965. *Insect Hormones.* Endeavour, Vol. 24, pp. 21–26.

Williams, Carroll M. 1958. *The Juvenile Hormone.* Scientific American, Vol. 198, pp. 67–74.

———. 1967. *Third-Generation Pesticides.* Scientific American, Vol. 217, pp. 13–17.

Chapter 10 Year of the Locust

Kennedy, J. S. 1956. *Phase Transformation in Locust Biology.* Biological Reviews of the Cambridge Philosophical Society, Vol. 31, pp. 349–370.

Uvarov, B. P. 1928. *Locusts and Grasshoppers: A Handbook for Their Study and Control.* Imperial Bureau of Entomology, London. 352 pp.

———. 1966. *Grasshoppers and Locusts: A Handbook of General Acridology.* Vol. I. Cambridge University Press, Cambridge, England. 481 pp.

Williams, C. B. 1958. *Insect Migration.* The New Naturalist; Collins, London. 235 pp.

Chapter 11 Parasitic Wasps, and How They Made Peyton Place Possible

Clausen, C. P. 1940. *Entomophagous Insects.* McGraw-Hill Book Company, New York. 688 pp.

DeBach, Paul, editor. 1964. *Biological Control of Insect Pests and Weeds.* Reinhold Publishing Company, New York. 844 pp.

Evans, H. E. 1963. *Wasp Farm.* Natural History Press, New York. 178 pp.
Swan, Lester. 1964. *Beneficial Insects.* Harper & Row, New York. 429 pp.

Chapter 12 Of Springs, Silent and Otherwise

Commoner, Barry. 1966. *Science and Survival.* Viking Press, New York. 150 pp.

Elton, Charles S. 1958. *The Ecology of Invasions by Animals and Plants.* Methuen and Company, London. 181 pp.

Farb, Peter. 1963. *Ecology.* Life Nature Library; Time, Inc., New York. 192 pp.

Laycock, George. 1966. *The Alien Animals.* Natural History Press, Garden City, New York. 240 pp.

Still, Henry. 1967. *The Dirty Animal.* Hawthorn Books, New York. 298 pp.

Chapter 13 Is Nature Necessary?

Bronowski, J. 1965. *Science and Human Values.* Revised edition. Harper Torchbooks, Harper & Row, New York. 119 pp.

Brown, Harrison. 1954. *The Challenge of Man's Future.* Viking Press, New York. 290 pp.

Darling, F. Fraser, and J. P. Milton, editors. 1966. *Future Environments of North America.* Natural History Press, Garden City, N.Y. 767 pp.

Eiseley, Loren. 1960. *The Firmament of Time.* Atheneum, New York. 183 pp.

Hocking, Brian. 1965. *Biology—or Oblivion: Lessons from the Ultimate Science.* Schenkman Publishing Company, Cambridge, Mass. 118 pp.

Platt, John R. 1966. *The Step to Man.* John Wiley and Sons, New York. 216 pp.

Index

Illustrations are indicated by italic folios